RADAR

· ·

The Silent Detector

These and other books are included in the
Encyclopedia of Discovery and Invention
series:

Airplanes: The Lure of Flight
Atoms: Building Blocks of Matter
Computers: Mechanical Minds
Gravity: The Universal Force
Lasers: Humanity's Magic Light
Printing Press: Ideas into Type
Radar: The Silent Detector
Television: Electronic Pictures

RADAR
The Silent Detector

by DEBORAH HITZEROTH

The ENCYCLOPEDIA of
D·I·S·C·O·V·E·R·Y
and **INVENTION**

Lucent Books P.O. Box 289011 SAN DIEGO, CA 92128-9011

HB FH WP

Copyright 1990 by Lucent Books, Inc., P.O. Box 289011,
San Diego, California, 92128-9011

Library of Congress Cataloging-in-Publication Data

Hitzeroth, Deborah, 1961–
    Radar: the silent detector/by Deborah Hitzeroth.

        p. cm.— (The Encyclopedia of discovery and invention)
    Includes bibliographical references.
    Summary: Examines the invention and development of radar, its
history, and its uses in science, law enforcement, navigation, space
travel, and the military.
    ISBN 1-56006-201-0
    1. Radar—Juvenile literature.   [1. Radar.]   I. Title.
II. Series.
TK6576.H58 1990   90-35500
621.3848—dc20   CIP
                                                        AC

15.95

Contents

■ ■

Foreword

The belief in progress has been one of the dominant forces in Western Civilization from the Scientific Revolution of the seventeenth century to the present. Embodied in the idea of progress is the conviction that each generation will be better off than the one that preceded it. Eventually, all peoples will benefit from and share in this better world. R. R. Palmer, in his *History of the Modern World*, calls this belief in progress "a kind of nonreligious faith that the conditions of human life" will continually improve as time goes on.

For over a thousand years prior to the seventeenth century, science had progressed little. Inquiry was largely discouraged, and experimentation, almost nonexistent. As a result, science became regressive and discovery was ignored. Benjamin Farrington, a historian of science, characterized it this way: "Science had failed to become a real force in the life of society. Instead there had arisen a conception of science as a cycle of liberal studies for a privileged minority. Science ceased to be a means of transforming the conditions of life." In short, had this intellectual climate continued, humanity's future world would have been little more than a clone of its past.

Fortunately, these circumstances were not destined to last. By the seventeenth and eighteenth centuries, Western society was undergoing radical and favorable changes. And the changes that occurred gave rise to the notion that progress was a real force urging civilization forward. Surpluses of consumer goods were replacing substandard living conditions in most of Western Europe. Rigid class systems were giving way to social mobility. In nations like France and the United States, the lofty principles of democracy and popular sovereignty were being painted in broad, gilded strokes over the fading canvasses of monarchy and despotism.

But more significant than these social, economic, and political changes, the new age witnessed a rebirth of science. Centuries of scientific stagnation began crumbling before a spirit of scientific inquiry that spawned undreamed of technological advances. And it was the discoveries and inventions of scores of men and women that fueled these new technologies, dramatically increasing the ability of humankind to control nature—and, many believed, eventually to guide it.

It is a truism of science and technology that the results derived from observation and experimentation are not finalities. They are part of a process. Each discovery is but one piece in a continuum bridging past and present and heralding an extraordinary future. The heroic age of the Scientific Revolution was simply a start. It laid a foundation upon which succeeding generations of imaginative thinkers could build. It kindled the belief that progress is possible as long as there were gifted men and women who would respond to society's needs. When Antonie van Leeuwenhoek observed *Animalcules* (little animals) through his high-powered microscope in 1683, the discovery did not end there. Others followed who would call these "little animals" bacteria and, in time, recognize their role in the process of health and disease. Robert Koch, a German bacteriologist and winner of the Nobel prize in Physiology and Medicine, was one of these men. Koch firmly established that bacteria are responsible for causing infectious diseases. He identified, among others, the causative organisms of anthrax and tuberculosis. Alexander Fleming, another Nobel Laureate, progressed still further in the quest to understand and control bacteria. In 1928, Fleming discovered penicillin, the antibiotic wonder drug. Penicillin, and the generations of antibiotics that succeeded it, have done more to prevent premature death than any other discovery in the history of humankind. And as civilization hastens toward the twenty-first century, most agree that the conquest of van Leeuwenhoek's "little animals" will continue.

The *Encyclopedia of Discovery and Invention* examines those discoveries and inventions that have had a sweeping impact on life and thought in the modern world. Each book explores the ideas that led to the invention or discovery, and, more importantly, how the world changed and continues to change because of it. The series also highlights the people behind the achievements—the unique men and women whose singular genius and rich imagination have altered the lives of everyone. Enhanced by photographs and clearly explained technical drawings, these books are comprehensive examinations of the building blocks of human progress.

RADAR

The Silent Detector

RADAR

Introduction

Radar is a device that locates objects by bouncing radio waves off of them. The term radar began as a code name during World War II and is an acronym that stands for *RA*dio *D*etecting *A*nd *R*anging.

Radar is used by the military, airlines, meteorologists, police, and astronauts. Without radar, soldiers could not watch for enemy airplanes; pilots could not avoid deadly thunderstorms; weather forecasters could not predict the weekend's weather; and astronauts would not be able to return safely to earth. But radar, which is so pervasive in our lives today, is a recent invention.

The story behind the development of radar is that of men and women looking for a way to protect themselves from an immediate threat. The first radar system was developed during the early years of World War II by Great Britain to protect itself from Germany's deadly planes. And, like many other technical fields, radar development has always advanced more quickly during times of war than times of peace.

For example, during the Korean War radar systems became much faster and

... TIMELINE: RADAR

1 | 2 | 3 | 4 | 5 | 6 | 7 | 8 | 9 | 10

1 ■ 1752
Benjamin Franklin proves that electricity can be controlled.

2 ■ 1819
Hans Christian Oersted makes the first link between electricity and magnetism.

3 ■ 1886
Heinrich Hertz becomes the first person to produce radio waves. During his research, Hertz also makes significant discoveries that lead to the later development of radar.

4 ■ 1894
Guglielmo Marconi discovers new way to communicate over long distances. Marconi links radio and telegraph technology to develop first wireless telegraph.

5 ■ 1921
Sir Robert Watson-Watt begins work on detecting storms by use of radio waves.

6 ■ 1922
Guglielmo Marconi develops a radar-type device to help ships avoid collisions in the fog. The device also is used to guide ships safely in to port.

7 ■ 1924
Edward Appleton bounces radio waves off the ionosphere. Through his experiments Appleton is able to measure the distance to the ionosphere.

8 ■ 1925
Gregory Breit and Merle Tuve, doing research for the United States Navy, prove the military uses of radar.

9 ■ 1934
British government asks Sir Robert Watson-Watt to develop a radio wave death-ray. Watson-Watt deems death-ray infeasible. Watson-Watt begins working on a radio wave detection system.

10 ■ 1935
Sir Robert Watson-Watt develops first radar system called a Radio Direction Finding System.

11 ■ 1936
A. Great Britain builds a series of early warning radar stations along its coast.

B. United States Navy begins installing radar aboard military ships.

more sensitive than systems used ten years earlier. This development was in response to the newer, faster jets being built for the war. Radar systems had to detect enemy aircraft at greater distances to give pilots enough time to intercept the enemy planes. Radar became even more complex during the 1970s and 1980s as the Cold War between the United States and the Soviet Union progressed. During this time, both countries began arming themselves with ballistic missiles that could be launched at land targets thousands of miles away. In response, both countries developed improved, computer-aided radar systems to defend themselves. High-speed computers helped the radar system detect and analyze the missiles to give the military more specific information about them.

While commercial development has lagged behind that of the military, it has progressed steadily. During the past fifty years, radar has been used for air traffic control, space landings and shuttle missions, and highway traffic control. In the future, radar may be used to search for disaster victims and may someday even guide us to our homes in outer space.

12 ■ 1940
Both Britain and the United States begin using radar as an offensive weapon. Radar is installed on U.S. and British planes.

13 ■ 1945
U.S. declassifies radar. Civilian radar research begins.

14 ■ 1949
Soviet Union tests its first atomic bomb. The United States begins looking for new ways to defend itself. In response, more powerful radar systems are developed.

15 ■ 1956
Collision between a TWA airplane and an United plane over the Grand Canyon leads to the development of a radar air traffic control system.

16 ■ 1961
Federal Aviation Agency rules that all commercial airplanes above a certain size be equipped with radar. The radar is used to detect weather and warn of approaching aircraft.

17 ■ 1974
National speed limit changes and police increase use of radar guns for speed enforcement.

18 ■ 1975
First radar detectors for motorists are sold.

19 ■ 1984
The National Aeronautic and Space Administration (NASA) begins using radar on board space shuttles to map areas of the earth's surface.

20 ■ 1986
Development is underway for Next Generation Weather Radar (NEXRAD), a system designed to gauge the size, intensity, wind speed, direction, and amount of water vapor in a storm.

21 ■ 1990
A. Plans are developed for *Freedom*, the National Aerospace Plane and space station.

B. Plans are developed for antistealth radar systems.

The History of Radar

During the early 1930s a violent civil war was being fought in Spain. Every night the radio was filled with stories of bloody battles and of the men fighting them. The stories often focused on Germany's airplanes and troops that were fighting with the Spanish troops led by Gen. Francisco Franco. Each night, Adolf Hitler and his Luftwaffe (air force) were becoming more infamous.

As the war progressed, the people of Europe, including Great Britain, began to fear Hitler and his lightning air force. The British were still trying to rebuild their country after World War I, a war that had killed a total of ten million troops

(top left) During World War II, German airmen prepare for assault against British forces.
(top right) Gunners dress in "swimming vests" in case they are forced into the English Channel.
(bottom left and right) A German plane takes off and heads toward the English shore.

German dive bombers in flight in 1940. British leaders were working on a device that would use radio waves—the main component of radar—to detect planes at a distance.

and wounded twenty-one million people. They did not need radio to tell them about the horrors of war. They had seen it for themselves when World War I had swept through Europe from 1914 to 1918. Now they were hearing stories that Hitler wanted to seek revenge on all of the countries—including Britain—that had helped defeat Germany during World War I.

British leaders had heard rumors about research being done with radio waves in Italy. They asked radio wave expert Sir Robert Watson-Watt about the possibility of using radio waves as a "death-ray" to shoot down Hitler's planes if they attacked. The death-ray they wanted to develop would have consisted of a high-intensity ray of radio waves. These waves would produce enough heat to kill the pilots of approaching planes. Such a ray was not possible, Watson-Watt told them. But, he suggested, it might be possible to use radio waves to detect German planes before they reached the English shore.

Scientists began working on a device that could detect planes at a distance. Within five years, the first radar system was developed. This system played a major role in helping Britain survive World War II and in helping the Allies win the war.

While radar seems to have developed quickly, the research that led to it had been going on for almost two hundred years. Before people could use radio waves, the main components of radar, to detect planes, they first had to learn how to create radio waves and to send them through the air. The development of radio waves was the result of experiments with electricity during the 1700s and 1800s. The history of radar is the story of scientists who were fascinated by the discovery of electricity and were determined to harness its energy for practical uses.

Lightning strikes

The first person to prove that electricity could be controlled was American inventor Benjamin Franklin. During his experiments, Franklin noticed that electric sparks looked and acted like lightning. After five years of experimenting with electric sparks, Franklin was convinced that

In 1752, Benjamin Franklin demonstrates his discovery that lightning is caused by electricity. With a kite during a rainstorm, he drew lightning down to the ground and established it as a form of electricity.

lightning was caused by electricity. He was sure he could harness electricity by drawing it down from the clouds to earth. Franklin was determined to test his theory, even if he had to risk his life to do it.

For the test, Franklin made a special kite out of silk. It had a wire projecting from the top of one of the sticks and a metal key attached to the end of the kite string. Once the kite was ready, all he had to do was wait for a thunderstorm. Finally, on a stormy June day in 1752, Franklin sent his kite aloft. His experiment was a success. He was able to make electricity created by the lightning storm travel through his kite and down the string to earth.

People who wrote about Franklin's experiment said he was like a mythic figure who could "draw down fire from the clouds" and "snatch lightning from the heavens." But Franklin had done more than that. He had proven that electricity could be controlled and conducted from one point to another.

Franklin wanted to prove that electricity could be put to a practical use, so he held a party for his friends. He planned to serve a turkey that had been killed by an "Electric shock, and roasted by the electrical Jack before a Fire kindled by the Electrified Bottle." Unfortunately his plan went awry, and Franklin discovered the dangers of electricity. In a letter to his brother John, Franklin described how close he came to dying that day:

> Being about to kill a turkey from the shock of two large glass jars ... I ... took the whole (charge) through my own arms and body.... The company

Inspired by the discoveries of Franklin, Hans Christian Oersted spent years uncovering the relationship between electricity and magnetism.

present say that the flash was very great and the crack as loud as a pistol: yet, my sense being instantly gone, I neither saw the one nor heard the other . . . . I felt what I know not well how to describe, a universal blow throughout my whole body . . . .

Franklin's experiments caused mixed reactions. Some scientists thought he was making major breakthroughs toward understanding and harnessing electricity. Other scientists and many of his friends thought his experiments were interesting but useless novelties. But Franklin knew his friends were wrong. He was sure that electricity would change the world. Franklin predicted a future where dinners would be cooked by electricity, wine glasses would be chilled by electricity, and guns would use an "electric battery." While Franklin was able to predict many of the wonders to come about because of electricity, even he did not foresee the invention of radar.

Wave of the future

One scientist who believed in Franklin's vision of the future was Hans Christian Oersted, a Danish professor of physics. While Franklin had proven that electricity could be conducted, Oersted would be the person who released its force on the world.

Oersted read all of Franklin's reports, including one in which Franklin said that electricity seemed to move a compass needle. Why, Oersted wondered, would electricity have any effect on a compass? Oersted spent the next twelve years trying to find out.

Oersted, a Danish professor of physics, paved the way for other scientists with his discovery that magnetic currents were created by electricity.

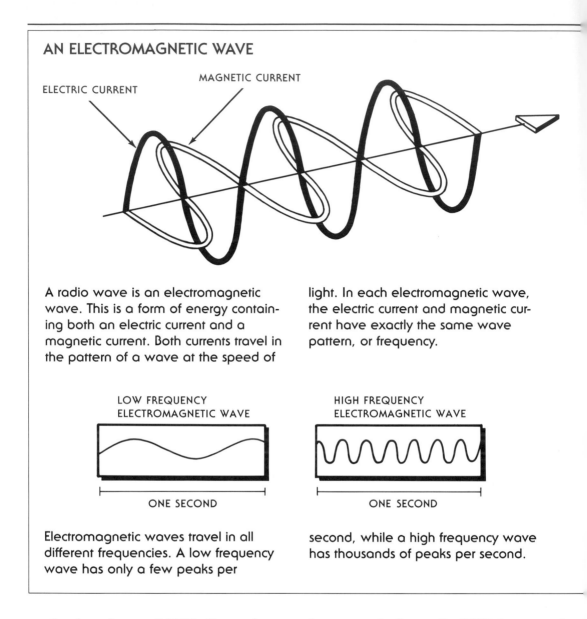

AN ELECTROMAGNETIC WAVE

ELECTRIC CURRENT

MAGNETIC CURRENT

A radio wave is an electromagnetic wave. This is a form of energy containing both an electric current and a magnetic current. Both currents travel in the pattern of a wave at the speed of light. In each electromagnetic wave, the electric current and magnetic current have exactly the same wave pattern, or frequency.

LOW FREQUENCY
ELECTROMAGNETIC WAVE

ONE SECOND

HIGH FREQUENCY
ELECTROMAGNETIC WAVE

ONE SECOND

Electromagnetic waves travel in all different frequencies. A low frequency wave has only a few peaks per second, while a high frequency wave has thousands of peaks per second.

In the winter of 1819, Oersted was performing an experiment for one of his physics classes. He noticed the needle of the compass he was working with wavered whenever he put an electric wire parallel to it. Even though he did not know what it was, Oersted was sure there was some force traveling from the electrical wire to the compass.

Oersted continued his experiments to unravel the mysteries of electricity and its magnetic force. By 1820 he proved that running an electrical current through a wire set up a magnetic field around the wire. Oersted's discovery was important. Once the relationship between electricity and magnetism was understood, people were able to develop magnetic devices to control and measure electricity.

What Oersted discovered was the basic principle behind radio waves. In

Invisible currents of electricity cross magnetic fields, creating colored light which shoots across the sky to create the arctic's dramatic aurora borealis, or northern lights.

very simple terms, Oersted and the scientists who came after him learned that an electric current produces a magnetic current and that a magnetic current creates an electric current. Radio waves are composed of alternating electric and magnetic currents. Electricity and magnetism interact to create a self-propagating wave that can travel through space. Initially, electrical energy is used to create a magnetic wave that in turn creates an electrical wave as it travels. The new electrical wave continues to travel through space and creates another magnetic wave which in turn creates more electricity. This explains how magnetism and electricity work together to create a wave of energy.

Oersted's discoveries also changed the way scientists saw the atmosphere. Before, it was seen as an empty place filled only with air. Afterwards, the world was viewed as a globe crisscrossed with invisible currents of electricity and magnetism. Some scientists speculated that

these forces might be responsible for many things that had been unexplainable before. One such was the aurora borealis, the magnificent display of colored light that streams through the sky in the arctic. Following Oersted's experiments, the aurora borealis was proven to be caused by magnetic and electrical forces.

James Clerk Maxwell was the first person to study Oersted's work and explain how waves of electricity and magnetism travel through the air. Maxwell's important theory laid the groundwork for other scientists to develop radio waves, the waves that are used in radar. As he developed his theory, Maxwell used the work of English chemist Michael Faraday and American physicist Joseph Henry.

Both Faraday and Henry had found that an electrical current moving through one wire could produce a current in another wire. This happened even if the two wires were not connected. Somehow, electricity was moving through the air to

James Clerk Maxwell developed the theory that electric and magnetic waves could be directed and received at a certain point.

He was content to develop ideas and let others try to prove them.

A stubborn man makes waves

Heinrich Rudolph Hertz was the man who proved Maxwell's theory by producing radio waves. Hertz performed his experiments because of a contest and an argument.

In 1879, Hertz was a twenty-seven-year-old physicist working in Berlin. During that year, the Berlin Academy of Science offered a prize for experimental work with electromagnetic forces. At the same time, Hertz's colleagues were arguing with him about Maxwell's theory. While Hertz believed it was possible to produce radio waves, many of his friends did not. They said that even though Maxwell could mathematically prove the existence of his

produce a current in the second wire. This was the first formal proof that there was some mysterious energy set in motion by electricity that could travel through the air. We now know that this mysterious force is magnetism. The electric current in the first wire sets up a magnetic current that flows to the second wire and produces a current of electricity.

Maxwell was the scientist who named the energy that travels through the air. When he presented his mathematical theory in 1879, he called this energy an electromagnetic wave. He considered this an appropriate name for the energy because it was made of both electricity and magnetism. Radio waves are one form of electromagnetic waves.

Maxwell established the theory that radio waves could be sent through the air and received at a distant point. Maxwell also thought that scientists could produce these waves in their laboratories. He did not, however, try to produce the electric waves he described. Maxwell was a mathematician, not a laboratory scientist.

Heinrich Rudolph Hertz modeled his work after the ideas of Maxwell. He proved that radio waves could be produced in laboratories.

waves, no one had been able to generate them. Hertz decided to take the challenge.

For eight years Hertz worked to produce Maxwell's waves, both to win the prize and to prove his friends wrong. Finally he was able to generate a spark of electrical energy that traveled through the air. He proved that Maxwell's waves existed. Hertz had found the secret of radio waves. He proved that they could be created, transmitted, and received.

Hertz also made discoveries that led to the development of radar. While working with the radio waves, he found that they would pass through all nonmetallic objects. Metallic objects, however, seemed to stop or reflect the waves. Hertz thought this was interesting, but he did not have equipment sensitive enough to detect and study the reflected waves.

With his experiments, Hertz started a revolution in the way scientists thought. Finally, scientists could produce and control radio waves, and they were eager to find a practical use for them.

While scientists all over the world were experimenting with radio waves, the first practical use for them was found not by a scientist but by a student.

A long distance call

Guglielmo Marconi, the inventor of the wireless radio, never claimed to be a scientist, but science was his obsession. Because his family had enough money to provide him with the best tutors and equipment available, he was able to make his scientific dreams a reality.

When Marconi first read about Hertz's experiments, he already was familiar with electricity. Marconi had been studying electricity since he was fourteen years old. He had been able to recreate some of the most famous electrical experiments at his father's estate in Bologna, Italy. Through his experiments, Marconi also became fascinated with the telegraph and the way it used electric signals to send messages across long distances.

In 1901, Guglielmo Marconi reads signals from an early tape recorder. He discovered a way to harness the waves created by Heinrich Hertz, and created the first wireless telegraph.

Because of his interest in the telegraph, Marconi's parents hired a retired telegraph operator as one of his tutors. From him Marconi was able to learn everything possible about the telegraph system.

In the telegraph system, operators sent electrical signals over wires stretched between two telegraph offices. The signals were sent by pressing a switch, called a key. This procedure sent electrical signals over the line in a pattern of short and long pulses. The electrical signals would travel to the end of the wire. There they would cause a magnet in the receiver to pull on a bar to produce clicks in the same pattern as the electrical signal. These "clicks" were Morse code, a system developed by inventor Samuel Morse to send messages through a combination of short and long sounds.

Even though the telegraph system was the best way to communicate across long distances at the time, there were many disadvantages. Experienced operators had to decipher the flood of clicks flowing out of the machine. And the telegraph really did not link the world; messages could only be sent between places that were connected by telegraph wires. Another drawback was that the system was not always dependable. Wind and rain could tear down lines, and lightning often interfered with the signal. Marconi was convinced there must be a better way than the telegraph system to communicate over long distances. He was looking for a long-distance communications system that did not require wires.

In 1894, during a summer vacation in the Italian Alps, Marconi was still considering this problem. While relaxing, he read an obituary on Heinrich Hertz in a

Samuel Morse holds a telegraph wire which carries electrical signals between two telegraph offices. He originated the Morse code, a coded series of long and short signals representing numbers, letters, and punctuation marks.

journal. The article described Hertz's work and discussed radio waves.

Marconi became excited. He was convinced that Hertz's waves were the answer to the problem of wireless communications. He thought that radio waves could carry electronic messages through the air without the use of telegraph wires. Years later, Marconi explained how he felt after reading the article. "The idea obsessed me more and more, and in those mountains of Biellese I worked it out in imagination."

When he returned home and told his friends and teachers about his idea for wireless telegraphy, many laughed at him. They told him that he should leave scientific discoveries to scientists.

Marconi refused to give up on his dream. He began working in his attic to learn everything he could about radio waves. First he built the equipment that Hertz described, then he recreated and improved upon Hertz's experiments.

Working in his attic laboratory, Marconi built a homemade radio wave transmitter. The transmitter produced a spark of electricity that was then broadcast through the air toward a receiver. Marconi used a curved metal reflector to direct the electric signal toward the receiver. The receiver was a sensitive instrument designed to listen for and receive radio waves. The receiver was hooked into a buzzer and a battery. Once the electrical signal was received, a device in the receiver would draw energy from the battery to make the buzzer sound.

Even though radio and radar technology today are much more complex, both systems still use the same components as Marconi's system. This system was equipped with a transmitter which produced the radio wave. It also had an antenna that broadcast the radio waves through the air and a receiver equipped with an antenna to detect the radio waves. Radio systems and radar systems are slightly different because they have different uses.

Radio broadcast systems send information and music across great distances, from the radio stations into radio receivers located in the homes of listeners. The radio sets have an antenna to detect radio waves. A receiver in the radio processes the waves from electronic signals into sounds so people can listen to music, weather, or news. But radar systems find objects by bouncing radar waves off them and listening for the reflected echo. To be able to listen for the reflected echo, radar sets are equipped with both a receiver and a transmitter.

One year after his first experiments, Marconi was able to transmit Morse code signals through the air for distances of three-quarters of a mile. This was the birth of the wireless telegraph. After his initial success, there was no stopping Marconi.

The radio age begins

In 1901, Marconi transmitted signals across two thousand miles of ocean. He had opened the door to the radio age, and the world was waiting to rush through. For years, scientists had been grappling with the idea of communicating over long distances without wires. Now Marconi, the person no one thought would succeed, had found the way to do it. Researchers were eager to try his experiment and improve upon it.

Some scientists carried on Marconi's work because they were challenged by his

The Italian-born inventor Guglielmo Marconi was able to transmit radar signals across the English Channel.

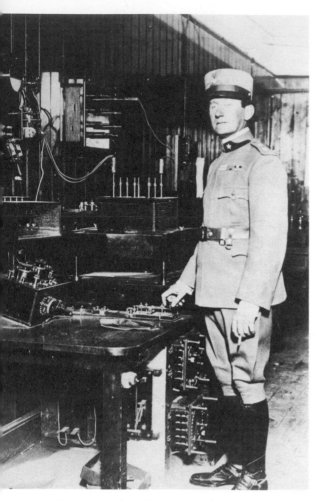
A man stands in front of a telegraph machine. Marconi discovered that signals could be transmitted not only by telegraph wires, but by radios as well.

wave beam came in contact with a metal object, it produced a hissing noise that could be heard in the receiver. Both Hertz and Maxwell had predicted that radio waves were reflected off metal objects. However, there had never been equipment sensitive enough to pick up these faint reflected waves. Now that Marconi had built a highly sensitive radio wave receiver, it was possible to detect the reflected waves.

When he was studying reflected radio waves, Marconi was working with simple principles. They are familiar to even young children today. Most people have shouted in an empty room or shower stall and heard their voices echo from a nearby wall. The echo is caused by sound waves bouncing off the wall and being reflected back. In this case, the person's mouth is the transmitter that sends the sound wave to the wall, and the person's ear is the receiver that detects the sound wave when it returns.

Marconi was interested in this effect, and he kept experimenting with it. He also shared the results of his experiments with other scientists.

In 1922 he gave a speech about his tests to a joint meeting of the Institute of Radio Engineers and the American Institute of Electrical Engineers. Marconi told the audience that he had "noticed the effects of reflection and deflection of these waves by metallic objects miles away." He also said that he thought reflected radio waves could be used at sea to detect ships through fog and rain. This would help avoid collisions. He closed the speech by predicting that the study of radio waves would "develop in many unexpected directions, and open up new fields of profitable research."

success. Others, hoping to become rich, continued to experiment with radio waves. For the first time, science was becoming a lucrative business; some inventors found themselves rich and famous almost overnight.

An echo of radar

While working with radio waves, Marconi made a discovery that opened the door for radar. He found that when a radio

War approaches

Marconi continued to experiment with reflected waves. By 1933 he had proof that metal objects could reflect enough radio waves for a receiver to detect. Marconi realized the potential military uses for such a device and contacted the Italian government.

Benito Mussolini, the dictator of Italy, read Marconi's reports. He was convinced that Marconi's experiments would play a major role in the war that everyone feared was coming. Mussolini ordered Marconi to conduct all future experiments in secret; no foreign assistants were allowed to help him. But Marconi's tests were difficult to conceal. Word soon spread that Marconi was experimenting with radio waves and aircraft. Not long after, rumors were circulated that Marconi was trying to develop a death-ray with radio waves.

These rumors reached Great Britain. Government representatives consulted Sir Robert Watson-Watt. They were looking for a death-ray to defend their country. The world was on the brink of World War II, and the stage was set for the development of radar.

Radar Is Developed

■ ■ ■ ■ ■ ■ ■ ■ ■ CHAPTER **2**

Edward Appleton was a pioneer in the development of radar. Appleton, a researcher at Cambridge University in England, was looking for ways to predict England's treacherous weather. He began his research after noticing that thunder and rainstorms caused static in radio broadcasts.

Appleton conducted experiments to find how weather affected radio waves. But, while performing meteorological experiments, he discovered that radio waves bounced off the ionosphere. The ionosphere is a layer of gas surrounding the earth near the top of the atmosphere.

Sir Edward Appleton, one of the pioneers of radar, received the 1947 Nobel prize for his experiments with radar and the atmosphere.

Appleton devoted much of his time to studying the atmosphere. Using the basic principles of radar established by Marconi, he measured the distance from earth to the ionosphere. Appleton transmitted short bursts of radio waves into the atmosphere. He then timed how long it took the waves to bounce off the ionosphere and travel back to earth. Using this method, he learned that the top of the atmosphere is sixty miles from earth. In 1947, Appleton received a Nobel prize in physics for these experiments. His measurement of the distance to the ionosphere was later viewed as the first practical use of radar.

Even though Appleton turned his attention from meteorology (the study of weather), research with radio waves and the weather did not stop. Other scientists were eager to continue his experiments. In 1915, Sir Robert Watson-Watt left the University of London to work for the Air Force Meteorological Office. There he hoped to do research on using radio waves to detect thunderstorms. But when he arrived at the office, Watson-Watt discovered other radio experts were already working on radio wave research. Instead of doing the work he had planned, Watson-Watt found his days filled with tedious tasks. Most days he worked with weather balloons and log books instead of with radio waves.

Even though he spent most of his time doing boring jobs, Watson-Watts' supervisors soon recognized his brilliance and determination.

Appleton tries to determine sun spot activity on radar by studying the sun through a telescope attached to an anti-aircraft radar director.

By 1917, he was meteorologist-in-charge of the office. In 1921, he was transferred to the Department of Scientific and Industrial Research where he could finally do the work he longed to do.

During the next fourteen years, Watson-Watt perfected his weather-tracking equipment. He completed more than six thousand experiments in a three-year period. These experiments allowed him to pinpoint the location of lightning and thunderstorms. During these years, Watson-Watt also developed the first display screen that roughly showed the position of the atmospheric disturbances he was tracking. This display screen was the same one he later used to build his first radar set.

Many of the experiments Watson-Watt was working on were also being conducted by other electrical scientists around the world. During the same time Watson-Watt was building his radio-tracking equipment, two U.S. scientists were experimenting with radio detection. A. Hoyt Taylor and Leo C. Young, working at the Naval Research Laboratory in Washington, D.C., conducted tests from 1922 to 1930. These tests indicated that radio waves might be of use to the military. But the U.S. scientists did not follow their experiments through to the point of actually inventing radar. Experimentation with radio detection in the United States was much slower than in Britain. The United States did not need to develop radar to defend itself because enemy planes had not yet threatened this country.

It was this need to defend their country during the impending war which sent the British government to Watson-Watt, asking for a radio wave death-ray.

Watson-Watt asked his assistant, Arnold Wilkins, to consider the problem. Wilkins later described his work on the Death Ray:

> I received the request on a piece of torn-off calendar. The request I soon deduced was to assess the fea-

Sir Robert Watson-Watt discovered that the advent of thunderstorms could be predicted with the use of radio waves.

sibility of a Death Ray, as he was asking me to calculate the amount of radio energy which must be radiated to raise the temperature of a man's blood to a fever heat at a certain distance.

Watson-Watt and Wilkins considered the idea for only half an hour before discarding it. They determined that the amount of energy needed for a death-ray would be impossible to generate. But during their calculations, they became convinced that radio waves could be used to detect the new metal planes being built in Germany and America.

By February 1935, Watson-Watt had developed plans for the first radar system, which he called a Radio Direction Finding System (RDF). The term radar was developed later by the United States Navy. The palindrome, a word which reads the same backwards or forwards, gives a clever hint of what radar does.

Like the experiments of Marconi, Watson-Watt's work was veiled in secrecy.

Working at the Naval Research Laboratory in Washington, D.C., Dr. Hoyt Taylor conducted experiments of value to military and naval operations.

Leo C. Young, an associate of A. Hoyt Taylor, experimented with military uses of radar.

During the first demonstration of his radar system, not even the pilot flying the test plane was told about the radar system. Flight Lieutenant Blucke was simply ordered to fly at an altitude of six thousand feet along a straight line between two British radio transmitters.

For the first test, Watson-Watt set up a mobile receiver inside his van and parked it one mile away from the transmitting towers. The purpose of the test was to see if the plane would reflect enough radio waves to detect when it passed between the two radio transmitters.

Tense and eager, Watson-Watt and his assistants waited in their van, staring at a small, blank, display screen. As the plane passed through the radio beam, a small blip appeared on the screen. The men watched as the blip slowly moved across the screen, mirroring the plane's flight through the radio waves. The test was a success. Britain had found its defense against Hitler's war machines.

In his dimly lit van that day in 1935, Watson-Watt and his small group of scientists created the technology that would detect planes. But this technology would also be put to other uses. It would guide pilots through dense thunderstorms, warn coastal dwellers of deadly hurricanes, and help people land on the moon.

Technology has made rapid advances since Watson-Watt's first experiment in 1935. Modern radar equipment is faster, stronger, smaller, and more efficient than it was in World War II. But the components used to track an F-14 plane as it blazes across the sky today are similar to the ones Watson-Watt used to track the single World War II Handley Page Heyford plane flying at only ninety miles per hour between the radio towers.

How radar works

All modern radar systems rely mainly on an antenna, transmitter, receiver, duplexer, and display screen. The antenna rotates on a vertical axis and scans the sky with a beam of radio waves generated by the transmitter. When the beam strikes an object, such as an airplane, radio waves bounce off the object and are sent back to the antenna. These waves are then sent by the duplexer to the receiver where they are amplified and displayed on the screen.

The principle behind radar is simple. The radar echo that returns to an antenna after being reflected from an object is similar to the echo that is reflected back to you when you shout in an empty room.

To further understand how radar works, imagine that instead of standing in a room, you are standing in a mountain valley. You would still hear an echo, but there would be a longer pause between the time you shouted and the time the echo returned. This is because the valley is much larger than the room. It would take longer for the sound waves your voice produces to travel to one side of the valley and back.

You can even calculate how far you are from the other side of the valley by the length of time it takes the echo to reach you. If you hear your echo four seconds after you shout, it means that it took two seconds for the sound waves to reach the other side of the valley and two seconds to return. Since sound waves travel at the speed of eleven hundred feet per second, you could calculate that you are twenty-two hundred feet from the other side of the valley. (1,100 feet per second times 2 seconds = 2,200 feet.) The experiment you have just imagined is similar to the way radar works.

But radar uses radio waves instead of sound waves because radio waves travel

This transmitter in Pie Town, New Mexico produces radio waves to create an echo which is reflected back to an antenna.

BASIC PARTS OF A RADAR SYSTEM

The OSCILLATOR generates the electric waves for the radar beam. From the oscillator, the electric waves are sent through the modulator to the transmitter.

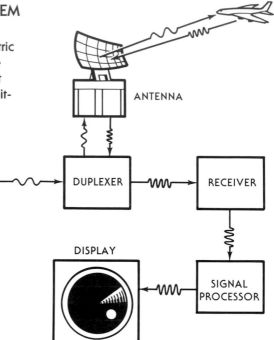

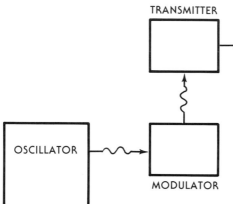

The MODULATOR acts as an on-off switch. It tells the transmitter when to send radar waves to the antenna and when to shut off.

The TRANSMITTER amplifies, or strengthens, the low-powered waves from the oscillator into high-powered electromagnetic waves. The transmitter usually generates short pulses of radio waves which last one-millionth of a second. It then shuts down for four milliseconds to allow the antenna to receive radio waves reflected from the target.

The ANTENNA receives signals from the transmitter and broadcasts them. After the transmitter shuts down, the antenna is used to receive the radar echoes reflecting from the target.

The DUPLEXER makes it possible to use only one antenna for both sending and receiving. It routes the radar waves from the transmitter to the antenna. When the antenna receives reflected radar waves, the duplexer routes them to the receiver.

The RECEIVER amplifies the weak reflected signals picked up by the antenna and sends them to the signal processor. It also filters out any background noise that is picked up by the antenna.

The SIGNAL PROCESSOR screens out echoes from large fixed objects, like trees or mountains, and sends only the desired signals to the display. Most modern radar systems use a computer as the signal processor.

The DISPLAY looks much like a television set. Most commonly, it shows a map-like picture of the area being scanned. The center of the picture corresponds to the radar antenna. Radar echoes are shown as bright spots on the screen. The distance of the spot from the center of the screen indicates how far away the object is.

faster, farther, and are reflected better than sound waves. Radio waves travel at the speed of light (186,000 miles per second). Therefore, radar waves can travel from a radar set to an object and return very quickly. A computer inside the radar set calculates the distance to an object in the same way you calculated the distance to the side of the valley. Radar calculations are performed by computers because the times involved are too short for a person to calculate quickly.

Once the computer calculates the position of the object, it is shown on the display screen.

Different radars do different jobs

There are two types of radar used today, continuous and pulse. The most common of these is pulse radar. In this case, the transmitter sends out short bursts of radio waves and then shuts off while the receiver listens for echoes.

Pulse radar is accurate in determining the distance and direction of an object. It is most often used in air traffic control, weather tracking, and in air defense.

Continuous-wave radar is used to determine the speed of a target. As the name suggests, radar waves are sent out continuously in this type of radar. Continuous-wave radar works on the Doppler Theory. This theory says that when radar waves bounce off a moving object their frequency will change.

To understand this, it is necessary to understand a little about radio frequencies. All radio waves are measured by their frequency. Frequency is defined as the number of waves that pass a point per second. Imagine you are on a beach watching a person standing in the water.

A man sits behind a screen in a Doppler radio room. Doppler radar measures the speed and location of an object.

WAVE FREQUENCY

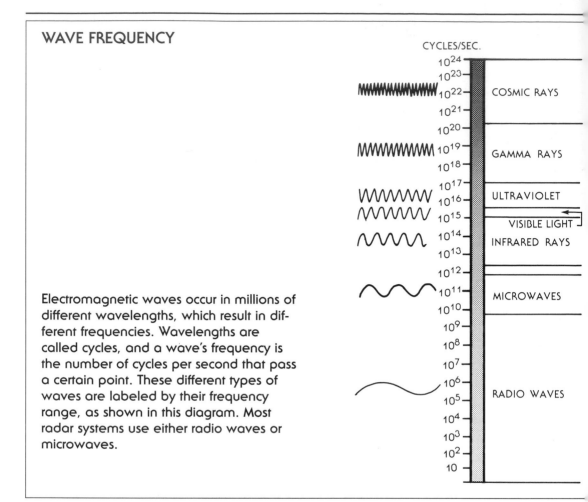

CYCLES/SEC.

Electromagnetic waves occur in millions of different wavelengths, which result in different frequencies. Wavelengths are called cycles, and a wave's frequency is the number of cycles per second that pass a certain point. These different types of waves are labeled by their frequency range, as shown in this diagram. Most radar systems use either radio waves or microwaves.

As you watch the waves rolling into shore, you can count the number of waves that pass the person in the water. The number of waves that you count is how frequently the waves come into shore, or the frequency of the waves. For example, if five waves pass the person in a minute, the wave frequency is five waves per minute.

Radar that measures the speed of an object is called Doppler radar. It works by sending out a continuous wave of radar beams. When the beam strikes a moving object, the reflected wave returns to the radar set at a different frequency. By measuring the difference in frequency between the transmitted beam and the reflected echo, the radar set can determine if the target is approaching or retreating. It can also determine the speed of the target.

This type of radar is frequently used by police to catch speeding motorists. It is also used by the military to determine the speed of a target.

Radar sets also are designed in different ways, depending on what the radar operator wants to look for. Some military radar sets have displays that show approaching targets only as blips of light. But other sets, those used by operators

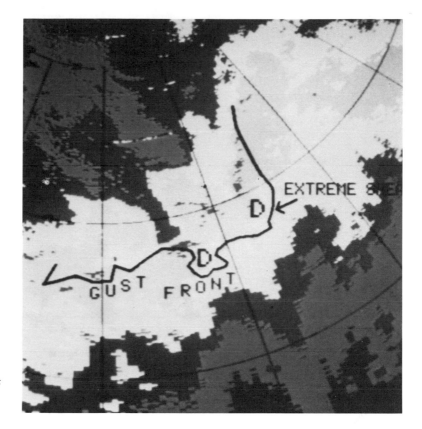

A weather radar display screen in Illinois produces an image of the line of a thunderstorm.

who need to know what type of aircraft or ship is approaching, have displays that show three-dimensional pictures of targets. Radar used by archaeologists display radar echoes based on the intensity of the returned signal. Wood, rock, and pottery all reflect radar waves differently. The radar computer can analyze each echo to determine if signals are being reflected by buried treasures or by layers of rock. Weather radar displays are designed to show wind and weather patterns as colored masses. In each case, the radar set is designed to fulfill the need of the person using it.

In times of need

Need has always been the cornerstone of radar. Even though the principle behind radar is simple, it took almost two centuries for it to be developed. It was not until Britain needed a new form of defense that scientists began experimenting with radar theory. And throughout World War II, radar development was linked to military needs. As the war progressed, radar was modified to meet new needs as they arose. Radar is still used more often by the military than by any other profession. The Army, Air Force, and Navy use radar to scan the skies for enemy aircraft, to track submarines under the ocean, and to follow foot patrols through rough terrain.

Radar and the Military

On August 15, 1940, Germany began its full-scale air attack on Great Britain. In the early afternoon, 170 German planes descended on the country. Thirty minutes later, another formation of fifty planes flew into view. The bombers continued to arrive until five hundred Nazi planes filled the skies. Even though the British were outgunned, they were able to shoot down 236 of the German planes and win the battle.

British pilots fought bravely that day. The real hero, however, was the radar system used to locate and track Hitler's

During World War II, a German plane drops bombs on England. Although the British were outnumbered, they were able to track down hostile aircraft with the use of radar, which played a major role in helping Britain survive the war.

If not for the use of a radar tracking system, the British would not have detected this German battle plane on its way to England.

air force during the fight. Each time the German air force crossed the British coast, a Royal Air Force squadron was there to meet it. The British planes were able to be in the right place at the right time because of Britain's early-warning radar system. With this battle, radar won its place in history and found a home with the military.

During World War II, radar became known as a "magic eye," which could pierce darkness, see through fog, and penetrate even the worst storms. Countries that possessed radar were able to carry out daring night raids and protect themselves from enemy bombers. During the opening years of the war, both sides were working frantically to discover and control the secret of the radar's magic eye.

Great Britain was one of the first countries to develop and use radar. By

1939, Britain had built a line of defense along its coast ranging from the Isle of Wight to the Firth of Forth. But other countries were soon to follow.

The radar race

The United States was slow to enter the radar race, but when it did it soon became a front runner. Following Japan's attack on Pearl Harbor in 1941, the United States put all of its resources into winning the war. The production of commercial television equipment was banned. Engineers working in the fledgling television industry were put to work on radar development. According to Michael Winship in his book *Television,*

> "We did not stop broadcasting immediately," [said one industry expert]. "What happened was that we were almost forced to stop, because we had very brilliant engineers, and the armed services needed their research talents far more than we did. So they reached in and took them."

The U.S. Navy ordered radar for its ships in 1936. Two years later, the battleship USS *New York* became the first ship to be outfitted with a full radar system. The United States was ready to fight the "Battle of the Beams," as the radar war was called.

But at the same time the United States and Great Britain were outfitting themselves with "magic eyes," so were their enemies. In November 1939, information smuggled to Allied forces gave a detailed description of Germany's research with radar. The papers confirmed that radar had been used to direct German fighters against British planes. They also confirmed that Germany was planning to use radar for night bombing missions.

(top)
Two women learn the mechanics of a radio hook-up. Operators were trained to use and maintain Britain's "secret weapon."

(bottom)
Recruits handle a secret radio-location plane detector which can locate hostile aircraft and direct anti-aircraft guns at the target.

These papers, which gave the Allies badly needed information about Germany's radar systems, came from a surprising source—a German scientist. According to Brian Johnson, in his book *The Secret War,*

> In the small hours of 5 November 1939 a parcel was left on the window-ledge of the British Consulate in Oslo, in what was still neutral Norway. Addressed to the Naval Attache, it contained several pages of German typescript... [these papers] set out the scope of German military scientific research ... [the papers] detailed German radar developments and confirmed that radar had been instrumental in directing fighters to a squadron of Wellington bombers that had been decimated on a raid on Willhelmshaven.... The letter was simply signed "A German Scientist who wishes you well."

At first many military leaders thought the papers had been planted by Hitler's men to mislead the Allied forces. But as the war progressed, the information proved to be true. In 1940, Germany began making night bombing runs against Britain. This confirmed that the information supplied by the anonymous scientist was reliable. Germany also possessed the magic of radar.

Radar changes the war

The invention of radar changed the way wars were fought. Before radar was developed, pilots who flew at night were flying

During World War II, mechanics test a radio-location unit. This device sends out sweeping radio-ray flashes which can detect ships, planes, and tanks.

A ground-control worker installs a radar system in an airplane. Armed with radar, a pilot can fly through fog, thunderstorms, and darkness.

"blind." They had no instruments to show them where they were going or what lay ahead. Without radar, pilots also were at the mercy of the weather because they had no way to spot storms and navigate around them. Ground-control personnel could do little to help. Without the help of their radio detection systems, ground control could not spot approaching aircraft or guide planes to safe air space.

But radar changed everything. From the beginning of the war, strong radar located on the ground was used to spot incoming planes and to direct pilots toward their targets. Toward the end of the war, after radar was installed in airplanes, aviators no longer feared thunderstorms, fog, or night flights. Instead of being life-threatening hazards, storms and darkness became the pilot's friend. They offered him cover while he slipped into his bombing target.

Radar also changed the way countries defended themselves. Before radar was developed, pilots would patrol large areas of the sky looking for approaching airplanes. The only way a country could stop an enemy bomber was for a patrolling pilot to visually spot the bomber and shoot it down. Squadrons would fly circuits around an area until they were low on fuel. When one air patrol was ready to return to base, another would be launched to take its place. Usually a patrol could stay in the air for about an hour before being forced to land and refuel. This constant patrolling took its toll in fuel, supplies, and in the readiness of the pilots for battle.

Radar made the air patrols unnecessary and introduced a new type of air warfare controlled from the ground. Instead of being patrolled by pilots, the skies were swept by powerful radar beams looking for enemy aircraft. From their positions in front of the radar scopes on the ground, radar operators could track enemy aircraft and direct pilots to their

During World War II, a five-man crew reads radar signals of an approaching plane. One tracks the direction of the aircraft, another tracks the elevation, and a third measures its range.

targets. Once an enemy patrol was sighted, the nearest fighter squadron was notified. Radar gave the pilots enough warning to take off and intercept the enemy aircraft before they reached their bombing targets.

In 1944, radar operators track enemy aircraft before notifying nearby fighter pilots to intercept the planes.

Even though radar was used extensively as an early warning device during the early years of the war, its full potential was not realized until the end of the war. Until the early 1940s, radar was used strictly as a defensive weapon to warn of incoming planes and to track aircraft during battle. But in the later years of the war, radar was installed on board planes and used as an attack weapon. The Royal Air Force used it extensively for night bombing raids. The United States Air Force used it to mark targets for daylight raids. As the war continued, the military realized they had found a versatile weapon in radar.

New uses for an old war hero

Following the end of World War II in 1945, military research with radar slowed. But as soon as the security ban on radar was lifted after the war, civilian scientists were eager to start radar research. During the next five years, scientists laid the groundwork for radar astronomy, radar navigation, and weather investigation.

In 1949, as the Soviet Union prepared to test its first atomic bomb, the military became interested in radar again. The United States found itself in need of a way to defend itself against an air attack. This was a position similar to the one Great Britain had faced before World War II. As the Soviet scientists continued their atomic tests, scientists in the United States began intensive research on new radar systems.

Contracts were given to both military and civilian companies to develop radar systems capable of tracking missiles and new high-speed planes. As radar devices became more complicated, computers were added to the systems to control and coordinate the devices. Radar also was linked with gun systems. This was to ensure that the radar computer could control the aiming and firing of guns and missiles aboard planes and ships.

All of the radar developments were made in response to the rapid advances in military technology. As planes became faster, radars needed to be able to track their positions and update them more frequently on the display screens. Radars also had to be more powerful to detect aircraft at greater distances. The faster the plane, the greater the need for early warning.

Some experts say the military "is addicted to radar." And looking at modern systems, it appears to be true. Fighter planes carry radar to spot enemy aircraft. Bombers use radar to locate ground targets and to fly at low altitudes. Cruisers and aircraft carriers are covered with radar systems. These systems also can be found on board tanks, on the backs of soldiers, and floating through space. This wide variety of radar systems can be clas-

When a navigator peers into a radar scope, he can see the terrain beneath him. Such devices guided bombers and squadrons to and from enemy territories.

sified in three areas: early warning, intelligence gathering, and weapons control.

Early warning systems

In response to the Soviet Union's atomic threat, the United States and Canada developed a ground-based early warning system to detect approaching aircraft. This system is called the Distant Early Warning (DEW) Line. It consists of fifty-two radar stations ranged across North America.

The DEW Line is composed of thirteen long-range radar stations and thirty-nine short-range stations. Information from the long-range radars is transmitted automatically by satellite to the North American Air Defense Commands Center. There the information is analyzed by radar experts. Information compiled by the short-range stations is initially analyzed by a computer. A synopsis of the data is then sent to the command centers.

Some opponents of the DEW Line system are urging that it be phased out and replaced with more sophisticated space-based radar. Critics of the ground-based system believe that it is possible for low-flying planes to evade the DEW Line and reach their target.

The development of ballistic missiles also created the need for a new, more powerful, radar system to detect the missiles which fly up to two hundred miles a minute. In response to this threat, the United States developed the Ballistic Missile Early Warning System (BMEWS, pronounced "bemuse"). BMEWS can detect missiles up to three thousand miles away.

BMEWS, like this one located in Thule, Greenland, was developed to detect powerful ballistic missiles before they reach their targets.

BMEWS consists of three powerful radar systems in Great Britain, Alaska, and Greenland. If a missile were launched at a target in the United States, BMEWS would compute its flight path and time of

BMEWS can detect missiles up to three thousand miles away that are flying up to two hundred miles a minute. This gives the U.S. a fifteen-to-thirty-minute warning of approaching foreign aircraft.

arrival quickly. The BMEWS would give the United States from fifteen to thirty minutes of warning to stop the incoming missiles, by exploding them before they enter United States air space.

The United States also has developed a space surveillance program. This program uses powerful radars to detect and track artificial satellites. This Space Surveillance System (SPASUR) collects data on hundreds of space-based objects. This includes reconnaissance satellites that are used for spying.

On a smaller scale, the United States also has developed personal radar systems small enough for soldiers to carry on their backs. These systems can detect moving tanks or trucks at distances of three miles. They can also detect a person crawling at distances of one hundred yards.

The Soviet Union also has developed radar systems to defend itself against enemy weapons. Soviet technicians have installed a sophisticated radar system at Krasnoyarsk, Siberia, to track ballistic missiles. In addition, they have developed radar and electronic oceanic-surveillance satellites to provide target information to vessels tracking enemy ships.

Weapons control

Radar also is used to test, control, and fire many military weapons. One of its most important jobs is missile guidance for many surface-to-air and air-to-air missile systems. The three types of radar missile-guidance systems used today are command, beam-rider, and homing. In command guidance systems, two ground-based radars are used to track both the target and the missile. The missile-guidance computers calculate the flight paths of both the missile and the target. Guid-

(top) A U.S. soldier carries a backpack personal radar system, which can detect surrounding trucks, tanks, and people.

(bottom) This Hawk missile guidance radar system is able to determine the path of a missile to its target.

THREE TYPES OF RADAR-GUIDED MISSILES

Using information from two ground-based radar systems, a computer continually calculates the paths of the target and the missile and determines their interception point. The computer conveys this data to a command transmitter, which steers the missile by radio signal toward the interception point.

COMMAND GUIDANCE SYSTEM

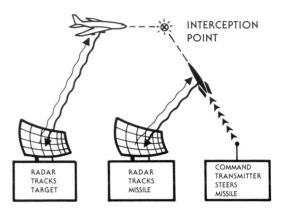

INTERCEPTION POINT

RADAR TRACKS TARGET

RADAR TRACKS MISSILE

COMMAND TRANSMITTER STEERS MISSILE

The missile contains a full miniature radar system and homes in on, or steers itself toward, the reflected signal from the target.

BEAM RIDER SYSTEM

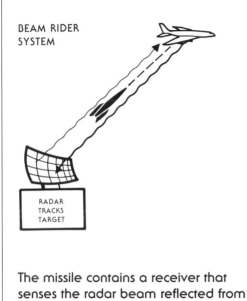

RADAR TRACKS TARGET

The missile contains a receiver that senses the radar beam reflected from the target. The missile follows the beam until it intercepts the target.

ance commands are sent by radio to steer the missile to intercept the target. In the beam-rider system, the missile is equipped with a radar receiver that senses the radar beam that is tracking the target. The missile changes course to stay within this beam and "ride" it to intercept the target. In the homing system, the missile is equipped with a miniature radar system. The missile transmits a radar signal that bounces off the target. It then "homes" in on the radar echo.

In addition to helping missiles find their targets, radar also is still used by pilots to mark their targets. Some planes, like the U.S. B-1 bomber, use radar to drop bombs. The B-1 uses radar to create an electronic bull's-eye on the spot where it wants to drop a bomb. Other planes, like the U.S. F-15, use radar in air combat

This mobile weapons guidance radar system assists pilots in locating enemy aircraft.

to mark approaching aircraft before firing missiles. Still other planes, like the U.S. Navy's S-3 Viking, use radar to find enemies hidden beneath the ocean. The S-3 can patrol over 2,300 nautical miles looking for enemy subs.

Intelligence gathering

Radar also is used to gather information about enemy activities and territory.

Military mapping planes are equipped with small radar devices that look downwards. As the plane flies over an area, the radar echoes are recorded on film instead of being displayed on a screen. The film is then fed to a computer, which produces a picture of the area that looks almost like a photograph. This radar system can take "pictures" of an area even when it is covered by clouds or darkness. This type of radar is used both for mapping terrain and for taking radar photos of enemy installations.

Other intelligence gathering systems include the U.S. Cobra Dane system, which observes Soviet ballistic missile tests. This sophisticated radar system is capable of following hundreds of missiles at one time as they spread through the sky.

A veteran of many wars

Since its creation, radar has been a valuable tool for the military. In World War II,

During air combat, this B-1 bomber can use radar to target the exact spot onto which it will drop a bomb.

This U.S. Navy S-3 Viking is equipped with a complete radar system to detect enemy submarines.

it guarded Britain's coast and guided pilots to their bombing targets. During the Korean War and the Vietnam conflict, radar was used to guide air-to-air missiles and to track supersonic jets. Radar has also been used by the Soviet Union and the United States to patrol the skies for ballistic missiles and enemy aircraft. If the present is any indication of the future, radar will be a serviceable weapon for years to come.

Radar and Navigation

On a dark, moonless night twenty-three hundred years ago, a silent Greek ship slipped out of the safe, well-known waters of the Mediterranean Sea and into history. The sailors on board were tense, eager, and scared.

The crew's first goal was to steal past the Phoenicians who guarded the Pillars of Hercules. If they were caught, they knew, their ship would be seized, and they would be sold into slavery. Some sailors were more afraid of what lay ahead of them than of the enemies who blocked their path. Once past the Pillars, now known as the Straits of Gibraltar, they would be sailing for uncharted seas.

Their captain was a young Greek man named Pytheas, a scholar, mathematician, and sailor. Pytheas was determined to prove his new method for navigating the sea.

Sailing from old to new

During the Fourth Century B.C., Greek seamen mostly sailed in the shelter of the protected Mediterranean Sea. Typically they sailed from port to port. They stretched their navigational skills only to avoid life-threatening squalls or their equally dangerous enemies, the Phoenicians.

When Pytheas began his great adventure, sailors navigated by piloting or dead reckoning. Piloting was the craft of navigating a vessel by following natural signs such as rock formations along the coast

or in waters close to land. For dead reckoning, sailors used known stars to estimate their direction, speed, and time of arrival.

Pytheas was convinced that people could locate places on earth by their relationship to objects in the sky. He had studied under the great Greek philosopher Aristotle, who talked about the earth as a great ball "suspended freely in space." Aristotle's lectures inspired him to test his theory.

During his journey, Pytheas sailed the coast of Europe, charted the waters

The Greek philosopher Aristotle inspired others to set sail into foreign waters and explore the world.

around Great Britain, and finally reached the Arctic Circle. During his voyage, he discovered people living "far to the north beyond any habitable land" and had many adventures. As he traveled a quarter of the way around the world, Pytheas laid the groundwork for celestial navigation. Celestial navigation is a form of navigation where sailors travel by following the stars based on their position in the sky at a certain time and date.

Ancient navigators became experts at reading the stars and the oceans. They learned to understand the ocean's tides, follow its currents, and anticipate its winds. They charted its depths by taking soundings with a lead weight tied to the end of a long rope. To measure the depth of the water, the rope was dropped over the side of the ship. The water depth was measured in fathoms, which originally was the distance of a man's outstretched arms. Today, fathoms are a standard six feet long.

Ancient mariners charted their courses carefully, and common trade routes were painstakingly mapped into the navigator's log. But all of their careful work could not protect them when fog or storms rolled in. The clouds blanketed the stars and the winds frequently blew them off course. Sailors needed a more reliable system for navigation.

In the early 1930s sailors finally found an instrument that would guide them safely into port, regardless of the weather. Guglielmo Marconi, the father of radio, was also an avid sailor. As his radio system progressed, he was eager to bring his miraculous invention to the aid of all seafarers.

Under Marconi's direction, the first radar navigation system was built. It consisted of two radio wave transmitters set at the mouth of a harbor. The transmitters were arranged so that their beams would overlap along an imaginary line. The line

A ship equipped with a radar system can detect and avoid obstacles in foreign waters.

THE RADIO DIRECTION FINDING SYSTEM (RDF)

Navigational radio waves, called radio beacons, are transmitted from navigational stations all over the world. Each station has a unique signal so that navigators can identify it on a map. When a boat's radar antenna receives one of these signals, it can determine exactly what direction the signal came from. Once a boat locates at least two radio beacons, its navigator can use simple geometry to figure out the boat's location.

led directly to the center of the harbor entrance. Ships were equipped with a receiver similar to a radio. If the ship wavered from the center line of the harbor entrance as it approached the harbor, the radar beams would produce a tone in the receiver. If the ship was too far to the left of the line, the beams would produce one tone. If the ship was too far to the right, the captain would hear a different sound. This system helped captains guide their ships safely into port. Marconi's system was similar to the Radio Direction Finding (RDF) system still in use today to guide ships.

The RDF system consists of a small boat-mounted receiver equipped with an extremely sensitive directional antenna. The antenna can be rotated 360 degrees to search for navigational radio waves. The beams, called radio beacons, are transmitted from navigational stations all over the world. In the United States these beacons are maintained by the U.S. Coast Guard. Once two or more beams are located, sailors can use them to chart their position in the ocean.

Most boats equipped for long-distance travel also have a full radar system on board. These systems transmit radio waves and listen for echoes, like all other radar systems. But instead of tracking enemy aircraft or searching for missiles, these echoes are looking for obstacles.

These include other vessels and shore lines, which can be hazardous to the unwary sailor.

Radar also helps ships find their way safely through calm bay waters filled with ships returning to port. When a ship enters a harbor, the vessel becomes the responsibility of the harbor master, whose job is to safely guide all of the ships pulling into or out of port. The harbor master uses radar to track each ship and to find the safest route through the harbor.

In addition to making navigation safer for ships, radar also has helped make air travel safer for planes.

An air tragedy sets new safety standards

Until 1956, air traffic control in the United States was mostly manual, and pilots relied on the principle of "see and be seen." According to this theory, pilots were supposed to avoid collisions by visually sighting other planes and maneuvering around them. Air traffic controllers helped prevent air collisions by keeping planes separated by "time and distance."

Manual air traffic control depended on the controllers' ability to plot a plane's position based on information supplied by the pilot. Planes were spaced in the air by time and distance. Controllers would command pilots to change speed or altitudes based on the distance they were from other planes. For planes to be accurately spaced, pilots were required to report their speed, altitude, and position throughout their flight. Air traffic control personnel used this information to calculate the approximate position of each plane and to determine safe flight paths. But the job was time consuming, and there was room for human error.

The Washington National Airport tower utilizes radar and computers to control air traffic. Radar is a main component of air control.

On the morning of June 30, 1956, the possibility for error in the system became a reality with tragic results.

At 9:01 a.m., TWA Flight 2 left Los Angeles headed for Kansas City. Three minutes later, United Flight 718 left Los Angeles headed for Chicago.

During the first twenty minutes of the flight, TWA 2 changed from its original altitude of nineteen thousand feet to twenty-one thousand feet, the same altitude as the United plane. The change was approved by the Los Angeles Center. Meanwhile, the United flight, flying a different course than the TWA plane, came under control of the air traffic control center in Needles, California.

TWA 2 reported to its flight center that it would reach the Painted Desert, Arizona, flight center location at 10:31 a.m. At 9:58 a.m., the United crew notified the communications station at Needles of its arrival time at the Painted Desert, also at 10:31 a.m.

On June 30, 1956, before the use of radar in commercial flight navigation, human error proved deadly when two airplanes collided with one another over the Grand Canyon.

At 10:31 a.m. an unidentified radio message was heard at the air control center in Salt Lake City. The message was: "Salt Lake, United 718 . . . ah . . . we're going in."

Military helicopters and civilian pilots later found wreckage of the two planes in the Grand Canyon. There were no survivors. The crash claimed the lives of all 128 people aboard the planes.

The Civil Aeronautics Board later determined that "the probable cause of this midair collision was that the pilots did not see each other." The report cited clouds, visual limitations of the cockpit, and a lack of air traffic advisory information.

Following the tragedy, a new air traffic control system was developed. This system called for an extensive use of radar and computers. In the more than thirty years since the Grand Canyon collision, radar has become an indispensable part of air navigation.

As pilots fly across the oceans of air, they are guided by radar beams from the ground. Ground stations transmit radar

The dotted line shows the strenuous route which Swiss and American mountain climbers took to reach the aircraft wreckage.

This radar screen in Stapleton Airport in Colorado is monitored by an air traffic controller.

through their air space. The first, and oldest system to be used, is called primary radar. It involves signals generated from the ground and reflected off aircraft. Echoes of approaching aircraft are picked up, processed, and sent to displays in the air traffic control center. Most controllers use video map displays that show towns, airports, and airways. All of the planes detected by the radar are shown as moving blips on the display screen.

Controllers also use secondary radar. This gives them more in-depth information than that provided by the primary system. It uses signals generated from aircraft and sent to the radar receiver. The information received from the airplane provides information on ground speed,

waves which flow out from the antenna in all directions. Pilots lock their receivers on to these beams and use them to navigate across the sky. In larger planes, these beams can be fed directly into automatic flight-control equipment.

Pilots also use radar to detect storms. By monitoring their radar system, pilots can detect rain showers and thunderstorms early enough to avoid the hazard.

All major airports also are equipped with radar systems to direct the continuous flow of planes taking off and landing. Most airport radars show the location of all planes in at least a fifty-mile radius. This helps air traffic controllers to direct planes along the safest altitude and path to avoid air collisions.

Air traffic control centers use two types of radar to track planes flying

The Dulles Airport in Washington, D.C. is equipped with a sophisticated radar system, as are most major airports.

AIR TRAFFIC CONTROLLER'S DISPLAY SCREEN

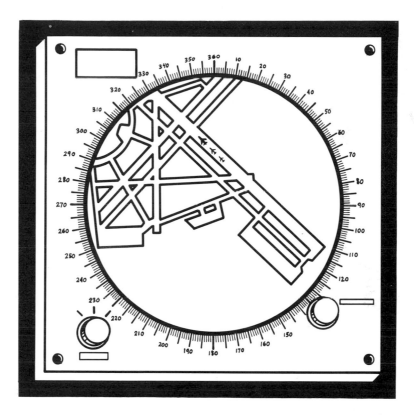

An object that is moving, either on the ground or in the air, will reflect radar signals from a slightly different location every time the radar beam strikes it. On the display screen, this will appear as several spots of light. On some display screens, the spots appear as images. The screen itself may contain a simulated map or picture to help the user determine the location of objects it tracks. The display screen above shows the movement of an airplane on a simulated map of a runway.

altitude, and the aircraft's call sign. This information is processed by the radar computer and is displayed adjacent to the aircraft blip on the display screen.

The use of computers and radar has made flying both safer and more efficient. Today most airports separate small planes flying at the same altitude by a distance of 1.5 miles; larger planes are spaced at distances of up to 6 miles. Without radar systems, small planes would need to be spaced up to 6 miles apart, and large jets might be as far as 20 miles apart.

Radar also has provided airline passengers with a smoother ride. Thanks to air radar, pilots can spot approaching thunderstorms and navigate around them.

(top) The first radar equipped air traffic control tower in the U.S.

(bottom left) Peering into a scope which gives a radar picture, an air traffic controller directs the landing of a plane.

(bottom right) An advanced automated workstation can process radar information and warn controllers of potential conflicts.

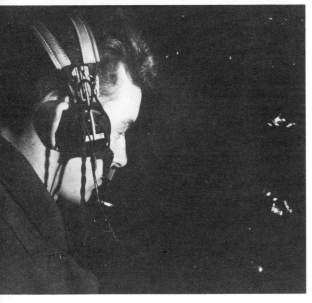

A safe port in the storm

Radar has revolutionized navigation on both the air and sea. Because of radar, men and women can safely navigate through unknown waters and arrive in foreign harbors unharmed. Radar also helps pilots maneuver their planes through the skies without fear of collisions or storms.

Radar and Science

Radar was Great Britain's greatest defense against enemy aircraft during World War II, but it was not infallible. Sometimes radar's magic eye worked too well. Instead of detecting airplanes hidden inside storm clouds, the radar waves would reflect from raindrops inside the clouds. When this happened, the radar operator would see a mass of radar echoes on his screen that marked the position of the storm. Echoes coming from the plane would be lost amid the echoes coming from the storm. Under certain weather conditions, pilots could hide behind storm clouds as they flew toward their bombing target.

Storm echoes were a major problem for radar operators throughout World War II. But since that time, they have been beneficial to meteorologists, people who study the weather.

Following World War II, the U.S. government allowed radar to be employed for commercial uses. Meteorologists were quick to continue the weather research started before the war by Sir Robert Watson-Watt. Before the development of radar, meteorologists had to depend on readings taken from weather balloons. The balloons would be sent up each morning, filled with equipment to measure air pressure, temperature, and humidity. The balloons were the only way weather forecasters could reach the upper atmosphere to detect approaching storms.

While the balloons provided useful information, much of pre-war weather forecasting was still a matter of guesswork. Over the years, weathermen and women became adept at determining how storm patterns would move. They learned to make educated guesses about whether certain pressure systems would produce deadly tornadoes or hurricanes.

Radar becomes a meteorologist

With the introduction of radar, forecasters could accurately foretell five-day weather patterns. Radar made it possible for meteorologists to watch storms brewing and track them as they traveled around the world.

Radar enables meteorologists to predict weather patterns, such as the thunderstorm shown brewing on this radar screen.

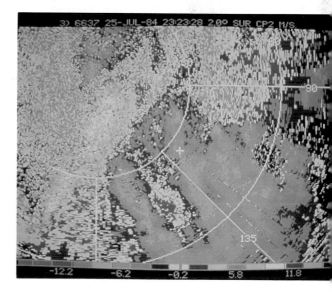

Lightning strikes near a radar dome. Tracking storms is especially important to pilots, who can be directed away from wind and ice.

The first weather radars were surplus military sets. The sets, originally used to detect and track enemy planes, were not ideally suited for weather use. This was because military sets were designed to track very large targets like planes, not something as small as a raindrop. These sets were used until 1951. At that time, the first set designed for weather detection was developed by the United States Air Force. This system was so sensitive it could detect even a light rainfall at distances of up to one hundred miles. A more modern system, the WSR-57, was later designed by the United States Weather Service to detect and track thunderstorms, tornadoes, and hurricanes. The WSR-57 did the same job as earlier models. However, it was faster and could detect approaching storms at even greater distances.

What is so special about weather radar?

Weather radar works much the same as any other radar system. Pulses of radar waves are sent out and received back by the radar set. But the major difference between weather sets and those used to track planes is the size of the radar waves. Radar experts have proven that radar systems with short wavelengths are better for finding small targets, especially targets as small as a raindrop. To understand this, we need a brief introduction to the concept of wavelengths.

Every wave has a high point and a low point. You can demonstrate wave motion by using a garden hose. Stretch the hose across the lawn in a straight line. Then, holding one end of the hose, quickly lift it and snap it back down as if you're cracking a whip. Notice the wave of motion that travels through the entire hose. When the wave reaches its highest point before going back down, this is called the top of the wave. When it reaches its lowest point before going back up, this is called the bottom of the wave.

Every radar wave, no matter how big or small, follows the same pattern as the hose. Each wave goes through the cycle, from high point to low point back to high point. If you could freeze your hose in place as the wave of motion passed through it, you could measure the distance from the top of one wave to the top of the next. The distance between the top of one wave and the top of the next is called the wavelength.

If a wave has a long wavelength, the arcs up and down are very big, and there is a great distance between one wave and the next. Long wavelengths are able to pass around raindrops. To understand this, imagine watching stunt drivers maneuver their cars between orange traffic cones. If the cones are set up in a row, the drivers can steer through them by moving their cars in a large wavelike pattern. If the pattern is large enough, the drivers will be able to avoid all of the cones. But if the cars' steering wheels are fixed so that the drivers can only make very small turns, they will not be able to avoid the cones. Radar waves with short wavelengths are like cars with steering wheels locked to make only small turns. The wave arcs are too small to pass around raindrops. Instead of passing around the drop, the wave collides with it. When the wave hits the drop, it sends back an echo to the radar set.

Thunderstorms hide deadly dangers from pilots

One of the most important things that weather radar does is track thunderstorms. Even though thunderstorms may seem pleasant to those of us watching from a distance on the ground, the thick black clouds can hide deadly perils from pilots. Meteorologists using ground-based radar and pilots using weather radar on board their planes track thunderclouds. They are looking for heavy rains, air turbulence, and hailstones.

Thunderstorms cause major problems in the sky and on the ground. Every year, thunderstorms sweep across the midwestern United States, inflicting millions of dollars in property damage from torrential rains and flash floods. Radar provides an early warning system so that property owners can protect themselves from hail and floods. With enough warning, people can move their cars into

Flash floods can cause severe damage, but if they are detected beforehand by radar, the damage may be minimized and people may be evacuated.

shelter to avoid hail damage. Also, people in flood-prone areas can seek higher land or start sandbagging their homes.

But rain and hail, which cause so much devastation on earth, are only minor problems for pilots. Unlike early aviators, today's pilots, equipped with radar, can identify heavy rain-pockets in clouds and fly around or over them. Lightning, too, no longer presents a major problem for pilots. All modern aircraft are built with skins designed to conduct electricity away from the plane and to insulate the delicate instruments inside the plane. If lightning strikes a plane, the aircraft's superstructure protects it from damage.

The hazards that pilots fear most when flying through thunderclouds are wind and ice. And radar can help protect pilots from both dangers.

Wind currents inside thunderclouds are different from those surrounding the clouds. Updrafts and downdrafts can sweep through the clouds at a speed of one thousand feet per second. Planes caught in these drafts can be slammed up and down like a model airplane on a windy day. And passengers always know when their pilots have been caught in major air turbulence—their smooth flight quickly turns into a wild roller coaster ride.

Hail is even more dangerous than wind currents. These ice particles from inside the cold layers of thunderclouds can range in size from small particles to balls as large as three inches in diameter. Hailstones can do serious damage to airplanes. In the past, even Air Force jets have been forced to land if they have received severe damage from hailstones hidden in thunderclouds.

Radar waves can detect both wind and hail, allowing pilots to fly around dangerous areas.

Airborne radar takes off

Serious threats presented by thunderstorms first prompted commercial airlines to start using radar. The airlines began to test airborne weather radar in 1947. At that time, the only sets available to commercial airlines were surplus military sets. By the early 1960s, special weather radar sets had been developed for commercial planes. These sets were much smaller than the military sets and were designed to detect weather and not enemy planes.

In 1955, Henry Harrison, director of meteorology for United Airlines, started looking for ways to detect hail by using radar. Harrison noticed that clouds that

The sequence of a thunderstorm is shown on a radar screen. These storms were an important factor in prompting the use of radar by commercial airlines.

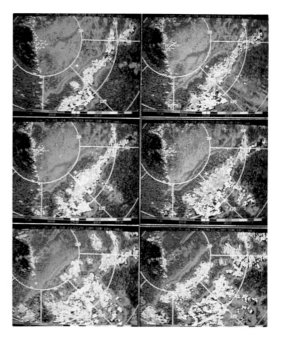

AIRBORNE RADAR

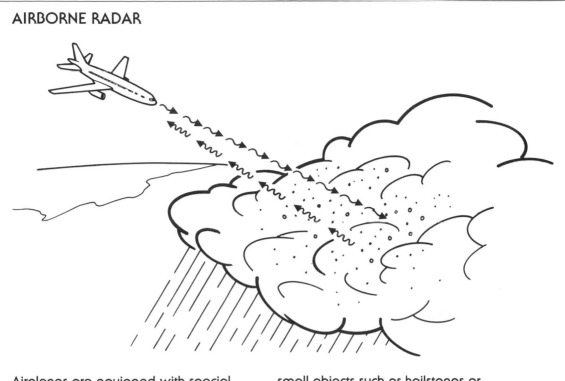

Airplanes are equipped with special radar systems that emit radio waves of extremely short wavelengths. These waves are effective for detecting very small objects such as hailstones or raindrops. Airborne radar allows pilots to detect storms early enough to steer around them.

were known to contain hailstones all returned a certain type of radar echo. In 1956, United ordered its pilots not to fly into clouds that returned radar echoes similar to the ones Harrison had found. Since that time, no United plane has suffered significant hail damage.

Other researchers later discovered more accurate ways to detect hail. Scientists at the University of Arizona and the University of Texas found that storms containing ice crystals reflected a stronger radar echo than those that contained only water.

In 1961, the Federal Aviation Agency ruled that all commercial airplanes above a certain size must be equipped with weather radar to aid in storm detection and avoidance.

Meteorologists show their claws

In 1982, meteorologists began experimenting with a new Doppler radar system designed to "see" the wind. The Classify, Locate, Avoid Windshear (CLAWS) radar detects the speed, direction, and patterns of the winds around an airport. Wind patterns are shown on the radar display in contrasting colors. The faster a wind is moving, the brighter its color on the display screen.

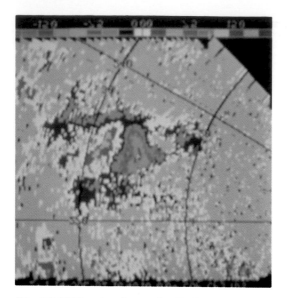

The CLAWS radar detects deadly downdrafts of air traveling at high speeds. Called microbursts, these downdrafts are extremely dangerous.

CLAWS was developed to help pilots avoid deadly microbursts during take off and landing. A microburst is a sudden, narrow downdraft of air traveling at speeds of up to 150 miles per hour. Microbursts were responsible for the 1975 Eastern Airlines crash in New York City, which killed 112 people; the 1982 Pan American World Airways crash in New Orleans, which killed 153; and the 1985 Delta Airlines crash in Dallas, which killed 133.

CLAWS patrols the area around an airport at ground level looking for deadly downdrafts. When one is spotted, the meteorologists contact air traffic control. Controllers steer pilots around the area, and advise them to avoid take-off or landing until the conditions improve.

Twister weather

While thunderstorms are the most common weather hazard, both on the ground and in the air, tornadoes are a much deadlier menace. Meteorologists must watch closely to see if a twister is building up inside a storm. Tornadoes consist of narrow columns of air flowing in circles. The wind speeds of the twisting mass of air can reach three hundred miles per hour.

The difficulty in predicting tornadoes is the speed with which they develop. While meteorologists can watch a normal storm build for up to five days in advance, tornadoes usually develop in less than an hour. And tornadoes can inflict millions of dollars of damage in only a few minutes.

To avoid tornado tragedies like the one that injured ninety-two people and caused $20 million in property damage

In 1947, this funnel-shaped tornado swept through Texas. New radar systems help to predict dangerous tornadoes.

In 1970, a Kansas Highway Patrol Trooper shot this remarkable sequence showing the formation of a tornado.

in Texas in 1986, a new, nationwide tornado warning system is being developed.

This system is called Next Generation Weather Radar (NEXRAD). It can gauge the size, intensity, wind speed, direction, and amount of water vapor in a storm. NEXRAD also can detect high-level, circular wind patterns that cause tornadoes.

The program, which is funded by the National Weather Service, the Federal Aviation Administration, and the Defense Department, calls for 160 stations to be built across the United States. Each station will cost approximately $2 million. This system will provide enough warning so people can take shelter and prepare their homes or businesses for the rapid pressure changes associated with tornadoes.

There she blows

When compared to a tornado, a hurricane is slow to develop. Hurricanes can take several days to build. Once they

NEXRAD measures and analyzes storm fronts and wind patterns, especially those that cause tornadoes, providing early warning to potential victims.

In 1989, the deadly Hurricane Hugo ripped through Charleston, South Carolina. Here, damaged boats litter the shore of an island off the coast of South Carolina.

arrive, however, they can be more devastating than a tornado. Hurricanes are made of violent winds reaching up to 150 miles per hour and torrential rains. While a tornado cuts a narrow path of destruction only one hundred yards wide, hurricanes can cover areas up to four hundred miles wide. Usually, the strongest winds cover an area of only fifty miles, but winds of up to seventy-five miles per hour can be felt through the rest of the storm system.

Hurricanes also cause major flooding, both from the heavy rains produced by the storm clouds and from giant ocean waves caused by strong winds. The major cause of death from a hurricane is drowning.

Strong weather radars located along the Atlantic Coast track tropical storms and monitor the buildup of hurricane conditions. As the hurricane develops, coastal residents are warned to prepare their property for high winds and to

move away from the shore. When hurricanes develop, many people flee coastal areas and travel inland.

Seeing through the trees

In addition to watching storms in the earth's atmosphere, radar also is used to study the earth's surface.

In 1984, the National Aeronautics and Space Administration (NASA) began using radar on board space shuttles to map areas of the earth's surface. Aerial photography had been carried out for years from satellites orbiting the earth. The photos, however, were useless in jungle or swamp areas. Masses of trees and other vegetation hid everything beneath them. NASA scientists knew that shuttle-borne radar sets could obtain clear pictures of what lay hidden beneath the trees.

Wrecked buildings and debris clutter the streets after Hurricane Hugo.

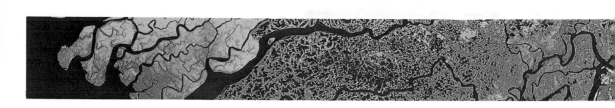

A picture taken from a shuttle radar set provides an image of the Ganges floodplain in Bangladesh. Scientists study pictures like these to learn about the environment.

The first shuttle radar project was carried out over the jungles of Bangladesh, near the Bay of Bengal. The project was led by NASA scientist Marc Imhoff. It was designed to pinpoint the location of stagnant pools of water that bred diseases such as malaria, cholera, dengue fever, and Rift Valley fever.

Locating and then treating or draining the stagnant pools would be major steps toward eradicating these diseases. The project would be especially helpful in eliminating malaria, which is transmitted by mosquitoes whose eggs hatch in stagnant water. In 1986, the World Health Organization declared malaria a major health problem throughout the world. An estimated 1.8 billion people were at risk of contracting the disease.

The experiment was a success. The space radar was able to penetrate the forests and find flooded areas of forest. Imhoff noticed the way the land was shaped. This allowed him to tell which areas would trap water and become stagnant pools as the flood water receded. This information helps scientists plan ways to drain pools or to build structures to prevent flooding in certain areas.

Looking beneath the surface

While Imhoff's group of scientists are interested in the earth's surface, others are interested in peeking beneath the surface. Radar that can penetrate underground is used today in engineering, transportation, and even in archaeology.

One of the most dramatic uses of modern technology is in the search for remains of ancient civilizations. Archaeologists are using radar aboard space satellites and airplanes to scan entire regions for possible excavation sites. On the ground, they are using truck-mounted and hand-held radar sensors to search for areas that might contain remnants of ancient civilizations.

Radar waves can penetrate earth, sand, and even volcanic ash, which cover the remains of other civilizations. When the waves strike rock, metal, pottery, or wooden objects, an echo is reflected back to the surface. By studying the number and intensity of underground echoes, archaeologists can determine the best location to dig for artifacts.

Radar was used to discover sites in areas such as Mount Hebron in the Middle East and the northwestern mountains of Costa Rica. In addition, spaceborne radar on board a space shuttle was used to find medieval settlements on the Baltic Sea. It was also used to look for early human settlements in Kenya. One of the most exciting finds was the discovery of ancient stream beds underneath the Sahara Desert. Scientists are following this lead to look for evidence of human habitation in the ancient desert.

Radar aids engineers

Archaeologists are using radar to excavate ancient engineering marvels. Other scientists are using radar to keep contemporary engineering wonders from being buried under a mound of dirt.

In New York City, thousands of commuters travel each day through Holland Tunnel. Made of brick laid inside steel forms, the tunnel was designed to last. But engineers who oversee repairs for the tunnel constantly check it for weak spots.

In the past they had only two ways to check the wall. These included doing manual soundings or measuring the water seeping through bricks. Neither method worked well enough. Manual sounding consists of tapping the walls and listening for echoes. This technique is used frequently in Europe. The practice, however, was deemed by U.S. engi-

A New York City tunnel is maintained with the use of radar. Here, motorists pay a toll as they exit.

neers to be too imprecise to ensure the safety of the cars passing daily through the tunnel. Checking water seepage gave a good indication of the general condition of the tunnel. It did not, however, pinpoint the location of weak spots.

To maintain the tunnel, engineers considered different methods to examine the thirty-two-inch-thick bricks. The system needed to be mobile and to be able to perform its test quickly. After considering different options, the engineers selected radar as the best system to use.

Mounted on a car or truck, the system can quickly scan the tunnel as it is driven through. This system has worked so well that New York City is considering using it to evaluate its subway system. Most subway tracks are set upon six to twelve inches of gravel above a four- to eighteen-inch-thick concrete pad. Some city officials are concerned that water seepage is eroding

As commuters travel through Holland Tunnel in New York, an engineer uses radar to check for weak spots in the tunnel structure.

Subway systems provide an important medium of transportation. New York City may soon use radar to scan the tunnel walls to ensure commuters' safety.

the dirt under the concrete. If the ground erodes too far, the concrete could shift and buckle the track. City officials are considering building a radar set that could be attached to a subway car. This set could scan the tracks and look for erosion before it caused any derailments.

In the future, radar sets mounted to subway cars will constantly monitor structure safety as they are driven through the tunnels.

Radar patrols the highways

Radar saves lives. It helps ensure public safety by guiding pilots safely through storms, by finding weak spots in tunnels, and by pinpointing possible defects in railroad tracks. Radar also saves lives by patrolling the nations highways, and helping police enforce traffic speed laws.

Radar and Law Enforcement

Law enforcement is probably the most well-known use of radar. Most people have either been in a car that was stopped by the police for speeding or know someone who has received a speeding ticket.

Most people understand that police use radar to find out who is speeding, but few people understand how radar works to catch speeders.

Police radar

To identify speeding vehicles, police use small Doppler radar units that can determine the speed of a target. Doppler radar units send out radar waves of a set frequency. All radar wave frequencies are measured in units called hertz. This unit was named after the German physicist who proved the existence of electromagnetic waves. If one wave were to pass a point each second, then the wave would have a frequency of one hertz. Police use two frequencies of radar, K-band and X-band. K-band radar operates at a frequency of twenty-four Gigahertz (one Gigahertz is equal to one billion hertz). X-band radar operates at a frequency of ten Gigahertz. Both types of radar units operate the same way for the patrol person.

Radar waves are sent out by a radar gun, which can either be hand-held or mounted on the dashboard of a patrol car. A radar gun resembles a pistol, except for its barrel. Instead of a slim barrel

like a pistol, a radar gun has a barrel horn, which looks like a megaphone. The horn contains an antenna that broadcasts the radar waves. The megaphone shape of the horn directs the waves forward and keeps the wave beam focused.

When the radar beam comes in contact with a target, it is reflected back to the radar gun. If the beam strikes a mov-

(top) A Connecticut police officer uses a radar gun to detect and catch speeding motorists. He is part of a squad known as the Bad News Bears.

(bottom) A motorist is unable to avoid the penalty for speeding.

DOPPLER RADAR

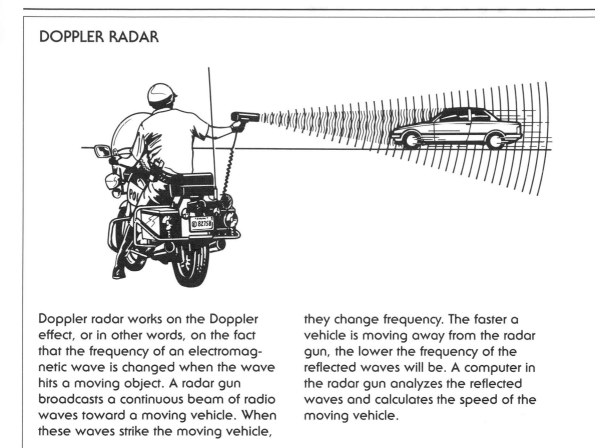

Doppler radar works on the Doppler effect, or in other words, on the fact that the frequency of an electromagnetic wave is changed when the wave hits a moving object. A radar gun broadcasts a continuous beam of radio waves toward a moving vehicle. When these waves strike the moving vehicle, they change frequency. The faster a vehicle is moving away from the radar gun, the lower the frequency of the reflected waves will be. A computer in the radar gun analyzes the reflected waves and calculates the speed of the moving vehicle.

ing target, like a car or truck, the reflected echo travels back to the gun at a different frequency than the transmitted wave. A computer inside the radar gun compares the frequency of the returned echo with that of the transmitted wave. After calculating the difference, the computer determines how fast the target was traveling to cause the shift in frequency. Once the calculations are complete, the speed of the target vehicle is shown on the radar display.

Radar guns can either be used in parked or moving patrol cars. Most radar guns in patrol cars have a split-screen display window, which shows both the speed of the target and the speed of the patrol vehicle.

Oil embargo creates need for radar guns

Radar guns came into wide use in 1974 when the national speed limit was lowered from seventy miles per hour to fifty-five miles per hour.

The speed limit originally was lowered during the energy crisis, which began in 1973 in an effort to save gasoline. The lower speed limit helped save gas because the faster a car travels, the more gas it uses per mile. The lower speed limit proved to have a double benefit: in addition to saving gas, the lower speed limit also saved lives. Following the speed limit change, the yearly highway death toll abruptly fell from fifty-five thousand deaths in 1973 to forty-six thousand in 1974. Eventually the United States was able to import more oil from other countries, and the gas shortage was ended. But officials retained the lower speed limit in an effort to reduce highway fatalities.

Since 1974, the fifty-five-miles-per-hour speed limit law has been broken frequently. One study in 1983 found that more than 50 percent of the traffic in thirty-seven states was exceeding the speed limit. Polls taken in 1984 asking drivers about their driving habits were even more revealing. Seventy-four percent of the drivers polled admitted to speeding occasionally. Another study of traffic on rural highways found that 75 percent of the people driving on the monitored roads were speeding. These statistics show why radar guns have become valuable to police across the United States. The radar guns are needed to detect the large number of motorists speeding on American highways.

The police departments' extensive use of radar guns has greatly increased the number of speeding tickets issued each year. During 1985, radar guns were used to help issue one million speeding tickets in the state of California. The same year, the state of New York issued 257,000 tickets.

A controversial history

But as popular as radar guns have become with police forces, they have never found favor with a large section of the American public. Opponents of radar guns claim that the devices are not accurate and that radar guns violate constitutional rights. In 1979, a group of Florida drivers took their case against radar guns to court.

A Maine State Trooper uses radar to check the speed of motorists on the Maine Turnpike.

This damsel in distress is actually a police officer, who, cleverly disguised, aims her radar gun at speeding motorists.

In the State of Florida vs. Aguilar and Consolidated Cases, eighty-one people contested the validity of tickets issued as a result of police use of radar guns. During the trial, radar experts tried to prove that radar guns were capable of clocking stationary "houses and trees and mailboxes" at speeds in excess of the posted speed limit.

Some of the testimony led to a sensational trial and made national headlines. Radar guns, however, were not proven to be inaccurate. What was uncovered during the trial was a lack of training by some police personnel who used radar guns.

Following the trial, studies found that most officers received only ninety minutes to two hours of training on the use of radar guns. One study also concluded that most training classes failed to teach officers about beam width or to explain the conditions that might interfere with radar beams. Both of these issues are significant because they can produce erroneous results.

Beam width is important because if the radar gun emits too wide a beam, it may cover multiple traffic lanes. If the gun is covering more than one lane, officers could receive readings from vehicles other than the ones they are trying to target. Studies have found that most radar gun beams spread far enough to cover two lanes of traffic at a distance of five hundred feet from the gun and four lanes at distances of one thousand feet.

While radar guns are usually reliable, they occasionally register erroneous readings. Radar guns display only the strongest radar echo they receive. And the strongest echo may not be coming from a specific vehicle that police are targeting.

For example, imagine a car traveling at the legal speed limit in the right lane of the highway. Then imagine a large truck traveling at seventy-five miles per hour in the left lane, two hundred yards behind the car. A police officer points his radar gun at the car and "shoots." As the wave travels, the beam spreads enough to encompass both lanes of traffic. The truck, which is a larger target, returns a much stronger signal to the radar unit. The radar gun picks up the strongest echo it receives and flashes the truck's seventy-five-miles-per-hour speed on its display screen. If the officer has not been properly trained to understand how the gun works, he or she might mistakenly issue a ticket to the driver of the car that was supposed to be the target.

There also are a number of conditions that can cause false readings. The presence of high-power lines near the radar unit, police radios, citizen band radios, chain-link fences, and power stations—all can have an impact on radar guns.

for officers. The Los Angeles Sheriff's Academy was a pioneer in radar training when it instituted a forty-hour radar usage course. Other states, including Florida and Utah, have followed Los Angeles' lead; they have begun their own forty-hour training classes.

Keeping a watchful eye on the watchful eye

While some drivers have chosen to battle radar guns in the courts, others are looking for methods to fight them on the highways. Not long after radar guns became widely used by police, drivers began to search for a way to avoid them. Within one year of the fifty-five-miles-per-hour speed limit, the first radar detector was introduced. Radar detectors are installed in cars and act as sensitive receivers that scan the air for radar waves. A radar detector continuously listens for radar waves. When it finds a radar wave in the same frequency that police use, the radar detector set emits a warning sound to alert the driver to slow down.

Radar detectors are effective because radar waves that are not reflected back to the radar gun continue traveling in a straight line after they leave the radar gun. The detector can pick up these stray waves up to a mile away from the radar unit. This usually gives the driver enough warning to slow down before coming into the radar gun's range.

One of the first radar detectors was produced in 1975 by Dale T. Smith, an electrical engineer who started his career producing radar guns for police departments. Smith's company, Electrolert Inc., built a radar detector called Fuzzbuster. The Fuzzbuster set off a controversy that is still raging today.

Radar guns, which can measure speeds of up to ninety-nine miles per hour, often result in traffic tickets for unwary motorists.

During testing of radar guns, *Motor Trend* magazine discovered that even small walkie-talkie radios can produce false results on the guns. When the walkie-talkies were in use, the gun gave false readings of speeds in excess of eighty miles per hour.

These were all factors discussed in the Florida court case. At the end of the trial, the Florida judge did not rule that radar guns were inaccurate. But he did rule that evidence from radar guns did not exclude all doubt.

In response to these findings, police forces across the United States have started special radar-gun training courses

Legal loopholes

Radar detectors are legal in all but two states and the District of Columbia. Legal organizations are attempting to outlaw them in all states. Since 1975, more than thirty bills have been introduced into state legislatures to outlaw radar detectors. All have been voted down, except for those in the states of Virginia, Connecticut, and the District of Columbia.

Opponents of radar detectors argue that the devices should be outlawed because their only purpose is to allow drivers to break the law. But so far, radar detector advocates have won most of the legal battles. Detector supporters claim that their right to use radar detectors is protected under federal laws. Supporters claim that the airwaves are covered under the federal broadcast laws. They argue that states do not have a right to make laws that curtail their use of the airwaves. With arguments like these, proponents of radar detectors have been able to keep them legal in most states.

One reason why champions of detectors are able to keep them legal is because radar detectors make up a large industry with a lot of money to spend defending its right to exist. In 1988, radar detectors were a $200-million-a-year industry. And the industry seems to be growing each year. Cincinnati Microwave, one of the industry leaders, sold 400,000 units in 1985. This figure was a 46 percent increase over the company's 1984 sales. And the industry is willing to spend a large portion of its profits on protecting its right to sell radar detectors. Electrolert, Inc. spent $3 million over a nine-year period to ensure that detectors remain legal. Companies that sell radar detectors finance public campaigns to keep radar detectors legal. They also pay high prices for legal representation in court battles.

Another point in favor of detectors is their popularity. Many legislators are hesitant to outlaw an item that is favored by a majority of their constituents.

Radar detectors, a trucker's best friend

Radar detectors are popular with all types of drivers. When the Fuzzbuster first became available in 1975, merchants could not keep them in stock. Stores sold out as soon as a shipment arrived. But radar detectors always have been most popular with truck drivers. One survey found that 80 percent of the truck drivers polled use radar detectors.

For truck drivers, who make their living driving, the penalties for speeding are high. States vary widely on the cost of a speeding ticket; but the average ticket for exceeding the speed limit by ten miles per hour is between thirty and forty dollars. In some states, fines can range up to one hundred dollars for a ten-miles-per-hour offense. And truck drivers risk losing their drivers' licenses and their jobs if they accumulate too many speeding tickets.

In the past, it was easier for truck drivers to hide their out-of-state speeding tickets from the license bureau in their home states. Law enforcement agencies do not report speeding tickets to license bureaus outside their state. It has been the truck drivers' legal responsibility to report their own tickets to the license bureau. Under this system, it was easy for truck drivers to hide tickets. They simply failed to report them to the license bureau in their home states.

But now things are changing. By 1995, truck drivers will be required to obtain

Because truckers often drive under time constraints, many use various types of radar devices to detect and avoid the highway patrol.

one national driving license. When this happens, all speeding tickets will show on each truck driver's record. It will no longer be possible to hide a ticket.

But even the shift to a national license may not prevent truck drivers from speeding. According to trucking experts, drivers must speed to make a living. According to one driving instructor, truckers who drive within the speed limit only average eight dollars per hour. This is because many truck drivers are paid per load they deliver. The longer it takes to make a delivery, the less money they make per hour on the load. To make more money, drivers must either speed illegally or overload their trucks. For many drivers, speeding is the easiest solution.

Kirk Hankus, a veteran long-distance truck driver, says that truck drivers use citizen-band radios, police radio scanners, and radar detectors to avoid the penalties of speeding.

Drivers listen to police scanners to discover where radar patrols are set up. When officers report their positions to a dispatcher, they also report it to truck drivers monitoring the scanner frequencies. Citizen band radios are useful because when one driver spots a patrol car using radar, the driver often will broadcast the officer's location on his or her radio. This alerts other drivers to the presence of radar. But radar detectors are the truck drivers' best defense against radar guns.

"You can't avoid the radar guns," Hankus said. "But these devices give you enough warning so that you can be within the legal speed limit when you are picked up by the radar."

Smile, you are on candid camera

As radar detectors become more sophisticated, law enforcement agencies are looking for new ways to enforce the speed limit. One way that is being tested in two cities is photo radar.

A photo radar system is a combination camera, radar, and computer unit. The system is set up in an unmarked police car along the shoulder of a highway. As cars approach the unit, the radar gun automatically clocks their speed. When a car is exceeding the set speed limit, the camera snaps a photo of the car. The photo includes the driver's face and the car's front license plate. The computer unit stamps the date, time, speed, and location on the film. Later, law enforcement officials develop the film, match the license plate number with the vehicle's owner, and mail a speeding ticket to the owner.

Even though this system is still in the testing stage, drivers are already challenging its legality. Ticketed drivers in Pasadena, California, went to court and won their case against the photo radar system. The California judge ruled in the drivers' favor.

Another legal problem to be resolved is who is responsible for the speeding ticket. In most states, when a car is stopped, the person driving is responsible. But with the photo system, whomever the car is registered to is responsible for the ticket regardless of who was driving.

The city of Paradise Valley, Arizona, has passed a special ordinance to deal with the problem. The ordinance states that whoever a vehicle is registered to is responsible for it. This is the same principle most cities use for parking tickets. The owner of the vehicle is responsible for its parking tickets, regardless of who was driving the car at the time it was ticketed. A citizens' action group already is fighting Paradise Valley's ordinance on the grounds that it is unconstitutional.

Check and checkmate

Police are always looking for new ways to enforce speed limits. Each time law enforcement officials find a new radar-detection method, drivers find a new way of detecting the radar guns.

Some researchers already are discussing the development of a radar system that would use laser beams to target cars as well as radar systems built into highway overpasses. But even as these ideas are discussed, radar detector companies are designing new devices to detect radar. Law enforcement's use of radar for speed detection promises to be a battle between technologies well into the future.

Radar and Outer Space

Radar was a faithful sentry for the Allies throughout World War II. But one day during 1942, the chain of radar stations along Great Britain's coast became useless. The island's entire radar system was jammed by a flood of radar waves coming from an unknown source. The country was left without protection for a brief period of time until the torrent of waves stopped.

Finding the culprit

Scientists quickly began working on the puzzle of where the waves came from. They needed to know if Germany had a new weapon that could make the British radar system useless. The investigators soon found out that Germany was not the source of their radar problems. The entity jamming their radar signals came from farther away—much farther. In fact, the culprit was almost ninety-three million miles away from the English coast.

The scientists found that the waves of energy that temporarily disabled the British defense system came from the sun. Giant solar flares had sent a flood of electromagnetic waves toward earth. These waves, made of the same elements as radar waves—electricity and magnetism—drowned out the radar waves produced by Britain's defense system.

Solar flares are eruptions of gas and radiation from the surface of the sun. These flares occur because the surface of the sun is marked with areas where the temperature is lower than the surrounding region. These zones of lower temperature build up exceedingly strong magnetic fields. When too much energy builds up in this magnetic field, a flare of hot gas and radiation is released. A large flare can release the same amount of energy as ten million hydrogen bombs. The temperature of a flare often reaches twenty million degrees.

The solar flares that blanketed Great Britain's defense system were much larger than usual, and they produced a massive amount of radio waves. The radar waves emitted by the English stations were simply lost among the radio waves showering the earth from the solar eruption.

This image of a solar eruption on June 10, 1973 was obtained during the first manned mission to Skylab 2.

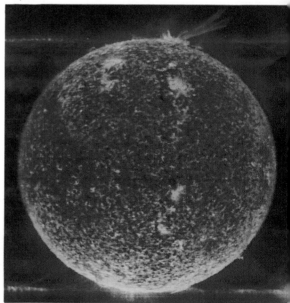

Comet Halley returns, rising over the Chilean Andes in the early morning of February 20, 1986.

Radar astronomy is born

It was this incident in 1942 that changed the field of radar astronomy from an interesting hobby into an actual science. Prior to that time, two scientists, Karl Jansky and Grote Reber, had proven that stars and planets emitted radio waves. But few people were interested in their findings.

One of the first people who thought scientists could learn about the stars by studying radio wave transmissions from the sky was an English physicist named Oliver Lodge.

Lodge was a pioneer in the field of radio communication. He first tried to detect radio waves from the sky in 1890, but he failed. His equipment was not strong enough to pick up the weak signals coming from the stars.

The first person to prove that radio waves came from space was an American radio engineer named Karl Jansky. In 1931, Jansky was investigating ways to reduce static in radio broadcasts. During his research, he detected all sources of static that were affecting his broadcasts—

except for one. This one source seemed to be coming from outer space.

Jansky published the results of his research in 1932 and 1933. But the radio industry was not interested in tracking the elusive cause of his static from space. Only one person, another American radio

The English physicist Oliver Lodge was the first to study stars and their radio wave transmissions.

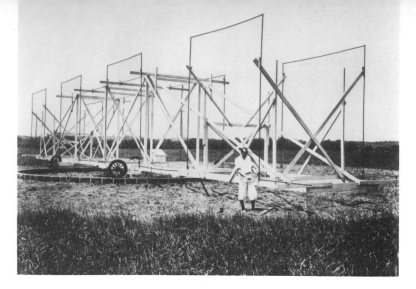

Karl Guthe Jansky made the first observations of radio waves coming from space with his rotating antenna system.

engineer, was interested enough in Jansky's reports to try.

Grote Reber read Jansky's report and decided to find the source of stellar radio interference. In 1937, Reber built a radio telescope designed to record the weak

The spiral Whirlpool Galaxy gives off emissions in the sky, which astronomers can plot on stellar maps.

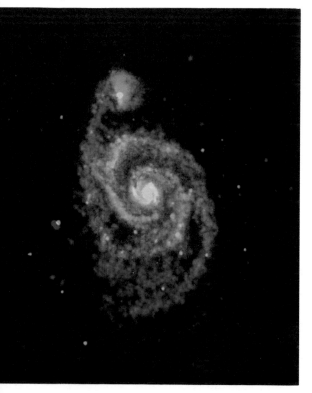

radio waves coming from the sky. Reber's telescope was actually a radio receiver thirty-one feet in diameter. Reber set up his telescope in his back yard. He then spent years carefully mapping the different sources of radio waves he found in the sky. Reber published his first report on radio astronomy in 1940, but few astronomers were interested in his experiments. Most people still thought that Reber's work was just an "interesting hobby."

The scientific world's interest was finally caught in 1942 when the solar flares drowned out Britain's radar system. Now scientists could no longer ignore stellar radio waves. Researchers began studying radio emissions from space to see what they could learn from the galactic radio.

Messages from outer space

Scientists learned much from quietly listening to radio waves that reached earth from the stars. Astronomers were able to mark the positions of various stars by plotting the intensity of their radio broadcasts on a stellar radio map. Following Reber's work, scientists found that the Milky Way gave off one of the strongest emissions in the sky. The Milky Way is a

Following Karl Jansky's work, Grote Reber built the first radio telescope and made radio maps of the sky.

After World War II ended, radar systems were available for public use. Astronomers then realized they no longer had to wait for radio waves from space. Now they could actively study the stars by sending out radar waves from earth and analyzing the returning echoes.

In 1945, scientists first used radar to study meteorite showers. Astronomers bounced radar waves off the meteorites to determine the speed and size of the objects in the shower.

In 1946, they went a step further and began using radar waves to study the moon. Radar waves can map its surfaces and help determine the exact distance of the moon from earth. In 1958, radar waves explored Venus, and in 1959 they probed the sun.

Radar gave scientists a new, accurate method for determining the distance to planets. Radar also allowed them to explore the size and shape of planets whose surfaces were hidden from view by clouds. Radar exploration brought back some surprising information.

The Reber telescope detected areas of strong radio emissions unrelated to any visible celestial body, proving that radar could be used in mapping unseen stellar bodies.

broad band of light that stretches across the night sky. The luminous band is composed of a vast collection of dim stars. Another source that broadcast strong radio signals was the Crab nebula. The Crab nebula is a crab-shaped mass of turbulent gas that looks like a misty cloud among the stars. The Crab nebula gives off a stronger radio wave signal than even our sun.

The Crab nebula is believed to have been created by a supernova in 1054. A supernova is an explosion of a star that causes it to lose between one-tenth and nine-tenths of its total mass. This violent explosion also emits a strong radio wave. Using radar astronomy, scientists have been able to mark the location of other supernova besides the Crab nebula. Scientists believe that three supernovas have occurred in our galaxy during the past one thousand years, in addition to the one that caused the Crab nebula. The supernovas happened in 1006, 1572, and 1604.

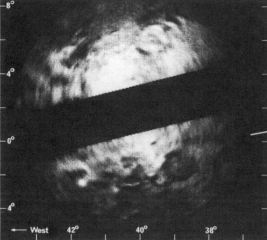

(left) The Crab nebula is a supernova remnant, seen on earth in the year A.D. 1054. It gets its name from its crab-shaped band of gases.

(below) At the Jet Propulsion Laboratory, radar astronomers have produced a map of cloud-shrouded, cratered Venus.

(bottom) A full view of the moon was photographed from the Apollo 11 *spacecraft during its journey homeward.*

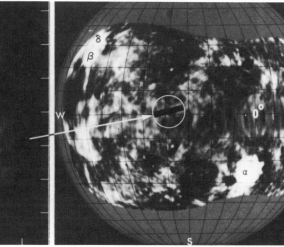

In 1962, astronomers discovered that Venus rotated clockwise. This was unexpected because all of the other planets in our solar system rotate counterclockwise. This means that if you were standing on the surface of Venus, you would be able to watch the sun rising in the west and setting in the east.

The tale of a comet

As space exploration grew during the 1970s and 1980s, astronomers were eager to put radar to new uses. Beginning in 1980, radar was used to detect and study comets as they blazed through the sky.

RADAR ASTRONOMY

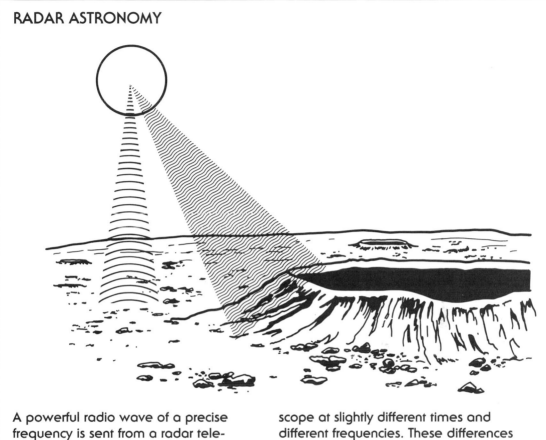

A powerful radio wave of a precise frequency is sent from a radar telescope toward several different spots on the moon, planet, or other object of interest. The echoes from these different spots arrive back at the telescope at slightly different times and different frequencies. These differences allow scientists to map the surface of the moon or planet being observed and make important observations about its composition.

Comets are celestial bodies that travel in extremely large orbits around the sun. When the comets come close enough to earth to be seen, they look like fuzzy balls of light streaking across the sky. Comets are harder to study than either planets or stars because they are smaller and return weaker radar echoes. But by using massive radar telescopes (some as large as one thousand feet in diameter) and radar systems, astronomers are learning about the composition of comets as they pass near earth.

The first one to be detected by radar was Comet Encke in 1980. Scientists also were able to detect Comet Grigg-Skjellerup in 1982, IRAS-Araki-Alcock and Sugano-Saigusa-Fujikawa in 1983, and Comet Halley in 1985.

One of the most successful radar studies came on May 11, 1983, when IRAS-Araki-Alcock passed 2.852 million miles from earth. Astronomers were able to probe the comet to find the size of its nucleus, or center, which is between three and ten miles. They also studied the rough surface of the comet. Scientists found that the comet's nucleus was sending out jets of gas. These caused a swarm of centimeter-size particles to extend 621 miles from the comet. The particle stream traveled for several days before scattering.

Halley was the most distant comet to be observed by radar. Scientists were able to detect the comet when it was fifty-seven million miles from earth.

Fifty-seven million miles is an impressive distance for viewing a comet. But the distance is small when compared to how far radar waves must travel to study the stars. In recent years, astronomers have used highly specialized radar telescopes and radar sets to send radar waves ninety-three million miles to study our closest star, the sun.

A closer look at the stars

In June 1989, radar telescopes and radar sets were used to record hydrogen movement in the sun during solar flares. Even though solar flares have been studied since 1942, scientists still know relatively little about them.

Scientists also use Doppler radar to study distant stars. Usually used by police to enforce speed limits, Doppler radar also is beginning to be used in other fields, including astronomy.

Astronomers use a radar device called a Doppler Imaging System that can construct a detailed radar picture of rotating stars. The Doppler system can detect areas of strong magnetic fields in a star, fields similar to the ones that cause solar flares. Doppler radar can track these magnetic fields as they move across the star. Most of the magnetic fields seem to start in the lower latitudes and eventually move toward the North and South Poles. Scien-

(top) Radar detected Halley's Comet fifty-seven million miles from earth.

(bottom) Intensely studied by astronomers when it came toward earth in 1986, Halley's Comet will next appear in 2061.

tists theorize that this movement is caused by the swift rotation of the stars.

Doppler radar also is being used to study the composition of stars. Researchers have found that some stars have an abnormally high level of metals as well as strong magnetic fields. Scientists are still working on theories to explain the erratic distribution of metals in some stars.

Astronomers are studying distant stars to learn more about the origin of our universe. By studying the bodies that are billions of miles away from us, scientists are hoping to unravel the mystery of how the universe began and how it has changed. As they find out more about the past, they hope to find clues to how the universe may change in the future. They hope the stars hold the answer to the question: What will the universe be like in one thousand years or in ten thousand?

While radar is helping earth-bound astronomers study the stars, it also is helping astronauts travel toward them.

Lift-off

Radar is vital to the space program. Without radar, astronauts would not be able to leave the launch pad or return safely to earth.

Ground-based radar is used to track spacecraft and rockets as they leave the launch pad and assume orbit above the earth. Following launch, radar stations around the world follow the craft's flight. This is done to measure the size and shape

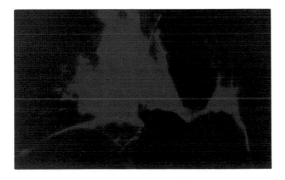

A solar flare explodes through the magnetic "fabric" of the sun, creating a ragged arch.

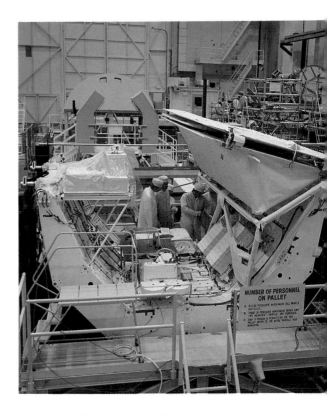

As spacecraft and rockets orbit above, they are monitored by ground-based radar stations.

of its orbit to ensure the craft is in its correct orbit and traveling safely.

Radar also is important for landing spacecraft. When American astronauts first landed on the moon, they brought radar with them. The specialized landing radar aboard the *Apollo* lunar module measured the distance from the craft to the moon as it descended.

As the landing craft approached its touchdown site, the radar system fed continuous information to the ship's computer and pilot. This information helped the astronauts regulate their speed as they approached the landing site. This was important because if the module had tried to land too fast, it might have crashed. If it was too slow, it would have wasted precious fuel needed for lift-off.

A starlit rendezvous

Once a craft is in space, the onboard radar system becomes the astronauts' guide. Radar is exceedingly important in helping astronauts rendezvous with other spacecraft. According to Michael Collins, an astronaut aboard the *Gemini 10* and *Apollo 11* missions, rendezvousing in space was "full of dark mysteries, and our experience as fighter pilots intercepting each other did not seem to apply."

When trying to dock, astronauts must consider the type of orbit of each craft, the speed of each vehicle, and the amount of fuel available for the intricate maneuvers. It is impossible for even the most experienced pilot to carry out the procedure manually. But radar makes the impossible job of a space rendezvous seem simple.

To determine the position, speed, and direction of the docking target, the spacecraft sends out a radar signal. Radar measures the speed and range of the target vehicle and sends the information to the pilot. This process is repeated twice to determine how fast the two craft are "closing," or coming together.

The astronauts use this information to approach the target vehicle the same way they would approach a moon-landing site. If the pilot approaches too fast, the spacecraft may pass the target, or even collide with it. If it approaches too slowly, the craft may have to waste precious fuel to accelerate to catch the target.

When the crafts are within one hundred feet of each other, the pilot's experience and skill take over. Much of the docking maneuver is handled manually and relies on the pilot's expertise.

Astronauts Neil Armstrong and Edwin Aldrin occupy this lunar module. In 1969, they landed and walked on the moon.

An eye in the sky

Radar also is used aboard spacecraft to study the earth. Like scientists who used ground-based radar to explore the skies, astronauts are using space-based radar to probe the earth.

For eight days during October 1984, the space shuttle *Challenger* carried an exceptionally sensitive radar system in orbit around earth. This system studied the world's atmosphere and environment. One of the crew's priority missions during the flight was to use the radar system to collect information on oceanography, hydrology (study of water distribution), forestry, geology, and archaeology.

This special radar consisted of a collapsible antenna equipped with two thirty-five-foot-long radar panels. As the shuttle orbited, the antenna was employed at three altitudes: 190 miles above earth, 148 miles, and 120 miles. During the eight-day mission, scientists collected only eight hours worth of data.

While this flight gleaned remarkable information about the world, some scientists were disappointed by the mission. Originally, the shuttle was to have collected forty hours of data. However, technical problems aboard the craft prevented the crew from performing all of the planned tests.

Part of the technical problems were due to the radar antenna itself. During the initial testing, the antenna jammed into open position. The shuttle could not fire its maneuvering system engines until the antenna's panels were closed for fear of damaging the delicate equipment.

From spacecraft, astronaut Sally Ride uses radar to explore the earth's atmosphere and environment.

Veteran astronaut Sally Ride was able to use the shuttle's mechanical arm to fold the antenna after each use. But valuable research time was lost each time the antenna was stored.

Other technical problems, including a malfunction with a relay satellite, caused the loss of the remaining thirty-two hours of radar research time.

Radar looks into the future

While studying the stars and earth, astronauts and astronomers have put radar to dramatic uses. But the future of radar promises even more spectacular developments. In the years to come, radar will be used for everything from guiding ships into space stations to locating the survivors of natural disasters like earthquakes.

The Future of Radar

On July 4, 1982, the space shuttle *Columbia* glided to rest on the Edwards Air Force Base runway in California. During the mission, radar helped guide the rocket that launched the shuttle. It also helped track the shuttle on its course around the earth and bring it safely home. As important as radar was to the mission, its role was small compared to what it may do in the future.

By the early part of the next century, astronauts may be using radar to guide their spacecraft home to a space station after a day of studying the stars. The space station will be a science station located in outer space. It will be designed to study the earth and stars. Plans are already underway to make this a reality. In 1990, the United States began planning the development of the National Aerospace Plane and Space Station, named *Freedom*. As plans progress, science fiction is quickly turning into science fact. And radar will be an integral part of the science of the future.

Military

The military always has been one of the chief users of radar, and it will continue to be throughout the next century. One of radar's most important military uses may be to combat stealth technology.

Stealth aircraft are designed to be invisible to radar. This is accomplished

The space shuttle Columbia *touches down at Edwards Air Force Base. Radar was used to launch and track the shuttle.*

Because of its shape and material composition, the Stealth Bomber is supposed to be undetectable by enemy radar.

through the shape of the aircraft and the use of advanced materials. Those materials include non-metallic substances (such as ceramics), composite materials (such as an epoxy-graphite mix), and radar-absorbent materials. In addition to being lightweight, these materials do not reflect radar beams as well as metallic materials do.

The shape of a stealth aircraft is just as important as the materials from which it is constructed. The aircraft is designed to diffuse incoming radar waves so that the searching radar detects less of an echo. Weapons carried by stealth aircraft are hidden inside the plane where they will not reflect radar waves. Everything is carefully designed so that the planes will not reflect radar waves.

According to top military officials, weapons to counter the new stealth technology will focus on radar. One of the most promising radars is ultra-wideband (UWB). UWB systems send out waves composed of thousands of different frequencies. UWB has several advantages over conventional radar in detecting stealth aircraft.

UWB provides a clearer picture on the display screen, which helps radar operators differentiate between targets. In addition to detecting stealth planes, UWB radar will help differentiate between incoming missiles and decoys. Decoy missiles are designed to give the same radar echo as real ones on traditional radar systems. But decoys are usually much smaller than armed missiles. UWB radar is able to detect and display the size difference; this allows radar operators to target only the real missiles.

UWB radar also is able to defeat the radar-absorbing materials that coat stealth planes. The materials are designed to absorb certain types of radar waves. Because UWB radar transmits such a wide range of frequencies, some of them are able to penetrate the coating.

If planning continues along current lines, these UWB systems may be linked together to form complex networks tied to advanced computer systems. The radar computers will possess as much as one thousand times the data processing power as existing systems. These computers will enhance the systems' ability to process even weak signals. With the help of the computer, the radar system will be able to turn these weak signals into three-dimensional images to help identify enemy targets.

The advanced technology necessary to build anti-stealth radars also will have added benefits for commercial industries. The technology used to build the complex computer-aided radar systems also will be applied to computer-aided production, automation, and robotics.

Friend or foe

In addition to helping spot invisible aircraft, radar will be used to differentiate between commercial and military aircraft. The tragic downing of an Iranian commercial airliner by the USS *Vincennes* in 1988 is a stark example of radar failure.

In this case, the radar system on the *Vincennes* misidentified the airliner as a military plane, and the crew of the *Vincennes* shot it down.

Since its inception in the early 1930s, radar has been designed as a detection system. When the British installed the first early warning system along their coast, they were interested only in detecting approaching aircraft. During World War II, there were no commercial airlines flying between enemy countries to worry about. And before the days of high-speed aircraft, pilots had time to make visual identification of approaching planes.

But today, as the sky becomes more crowded with commercial and military aircraft, detection is not enough. In addition to knowing where an aircraft is, radar operators need to know what it is. Is the plane a friend or foe? Is the radar blip tracking a commercial or military craft? These questions are vital to military commanders who must make split-second decisions to fire missiles or to allow the airplane to approach unharmed.

In 1988, radar failure on the USS Vincennes *caused the misidentification of an Iranian airliner. It was identified as a military plane and shot down.*

The new radar system, Traffic Alert Collision Avoidance System, will provide a more sophisticated way for pilots to detect nearby planes.

Two different programs are being researched to provide radar operators with the information they need.

One target-identification technique uses radar waves to create a two-dimensional image of the aircraft. To produce these images, the radar system uses a number of radar images taken as the plane moves across the sky. The images are combined electronically to produce a crude "snapshot" of the target plane. This process is amazingly fast. According to plans being tested, the system will be able to take radar images of the plane at a rate of four hundred per second. One disadvantage to this system is that radar operators have to be highly trained to identify the silhouette as a specific aircraft.

Another more promising system changes the radar echoes into colored waveforms on a computer screen. Operators simply match the waveform and color to charts to identify the type of approaching aircraft.

This system works by sending special radar pulses toward the aircraft and listening for the aircraft's vibrations. Research has found that radar pulses of certain frequencies will make a metallic object vibrate based on the shape and construction of the object. When the radar wave hits a plane, it causes the metal in the plane to vibrate like a giant tuning fork. Every type of aircraft vibrates uniquely depending on its size and shape.

The reflected echoes, which are in tune with the vibrations, are picked up by the radar antenna. The echoes are digitized and fed through the radar computer's memory. The computer matches the echo with information it has stored on the echo patterns of different types of aircraft. After matching the aircraft, the targets are displayed on a computer monitor.

One problem with this system is that in some cases the same body-types are used for both military and nonmilitary aircraft. Researchers are working to improve the system so that it can distinguish between two identical aircraft and even determine which one is carrying weapons.

A new coast guard

In addition to developing systems to identify planes, the military also is building new, traditional radar systems designed

The wide radar range of Raytheon's Relocatable Over-the-Horizon Radar System will give the U.S. Navy early warning of approaching aircraft and ships.

to detect approaching planes and ships. The U.S. Navy is working on a Relocatable Over-the-Horizon Radar (ROTHR) system. Over-the-horizon radars are designed to make use of an effect known as ducting. The radar waves get trapped in a layer of air and can travel farther than normal. These layers are created by certain temperature and pressure conditions in the atmosphere. Surface ducts can often be found within one hundred feet of the surface of an ocean. This allows the ROTHR to detect ships at much greater distances than normal radars because the waves travel farther. Even higher above the earth than surface ducts are elevated ducts. These ducts are like a tunnel in the air, varying in depth from one hundred feet to one thousand feet. Radar waves transmitted into these tunnels can detect aircraft at extended distances. The ROTHR system is currently being tested in Virginia and Alaska.

If testing goes as planned, the operational unit will be in place during the mid-1990s. This system is expected to play an important role in the anti-drug war because it can detect drug-smuggling planes entering the United States.

The test system in Virginia uses radar waves to cover a 1.6 million-square-nautical-mile region. This region extends from the coast of Colombia through Nicaragua, Honduras, and Florida, as well as from Puerto Rico to Venezuela. During initial testing at the ROTHR site in Whitehorse, Virginia, the system was able to detect several small aircraft believed to be involved in drug trafficking.

Air traffic control

In addition to detecting hostile aircraft and aircraft carrying drugs, radar of the future will be busy protecting friendly planes.

A new airborne radar system called Traffic Alert Collision Avoidance System (TCAS) is being developed to help pilots detect and avoid other aircraft.

"SEEING" OVER THE HORIZON WITH ROTHR

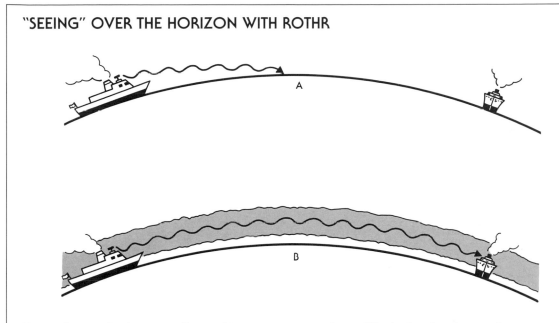

Normally a radar beam will travel in a straight path until it strikes a solid object. Since the earth's surface is curved, a beam that does not strike another object will eventually run into the horizon (A). The idea behind ROTHR (Relocatable Over-the-Horizon Radar) is to aim the radar beam so that it travels through a "duct" in the earth's atmosphere (B). A duct is a layer of warm, moist air that is more dense than the surrounding air. The density of this layer causes some radio waves to be trapped inside it. Then, instead of traveling in a straight path and bumping into the horizon, they follow the curved path of the duct around the horizon.

TCAS is an on-board radar system that plots the bearing and distance of other transponder-equipped airplanes flying nearby. A transponder is a device that transmits information about the plane. When a transponder is hit by radar, it automatically transmits encoded radio data about the plane. Transponder information includes the plane's code number, airline, and rate of ascent or descent.

TCAS is equipped with a radar system designed to interrogate transponder units. When a plane approaches, TCAS sends out a radar wave to activate the plane's transponder. An antenna in TCAS is designed to receive information broadcast by the transponder.

TCAS also includes a high-speed computer that uses information from its radar system and the transponders of approaching planes. The computer plots the closure rate and range of approaching planes. The information also is used to help pilots develop possible course changes to avoid potential collisions.

Like many other types of radar, TCAS will be equipped with a plane position indicator. This will display the TCAS-equipped plane at the center of the radar screen. Approaching aircraft will be shown as blips of light surrounding the TCAS craft. If an aircraft approaches too closely, or enters a flight path in conflict with the TCAS-equipped craft, the approaching plane is flagged as a threat.

To determine if a plane is becoming a threat, TCAS calculates factors. These include altitude, closure rates, airspeeds, and flight paths. As planes become threats, TCAS changes their color on the radar display screen. When a plane is tagged as a threat, TCAS will display information on ways to avoid the approaching plane.

Targeting deadly winds

While TCAS will help pilots avoid air collisions, another on-board radar system is being researched to help pilots avoid invisible dangers.

In the late 1980s, the National Aeronautics and Space Administration spent $24 million to fund a five-year program. The purpose of this program was to develop an airborne system to warn pilots of possible microbursts in clouds.

Studies in 1986-1987 found that even small, harmless-looking clouds could produce deadly microbursts.

At most major airports, systems are in place to warn pilots of microbursts during takeoff and landing. But there is no present system that can warn pilots of microbursts at higher altitudes.

Current research may develop a Doppler radar that can be reflected from snow, water droplets, hail, and even insects in the clouds. Such reflections would help to determine the direction and speed of wind movements in a cloud. The results of the study should be finished during the 1990s. An airborne system to detect microbursts may be developed some time in the early twenty-first century. Once the system is installed, pilots will be able to fly through clouds without fear of falling prey to deadly winds.

A friend indeed

Since its development in the 1930s, radar has been protecting people and saving lives. Initially developed to defend Great Britain's coast against Germany's attack, radar now guides pilots safely through thunderstorms and deadly winds. During the past sixty years, radar has helped navigators travel across uncharted seas, taken people to the moon and helped them return safely, and has helped reduce life-threatening illnesses in the jungle.

During its short history, radar also has enabled people to learn more about the universe. Radar has spotted comets as they stream across the sky, has mapped stars and probed planets that cannot be seen with a normal telescope, and has delved beneath the earth's surface.

And radar of the future seems to be headed in the same direction. As people develop new needs for safety, protection, and information, new radar systems will be there to guide the way.

Glossary

■ ■

antenna: the part of the radar set that receives signals from the transmitter and broadcasts them in a tight beam. After the transmitter shuts down, the antenna is used to receive the reflected radar echo.

ballistic missile: a self-propelled missile developed during the 1950s.

dead reckoning: navigating a ship by using the stars to estimate direction, velocity, and time of arrival.

display: the part of the radar set that shows where objects detected by radar are located. The display resembles a television set and usually shows a map-like picture of the area being scanned. The center of the picture corresponds to the location of the radar set. Radar echoes are shown as bright spots on the map.

ducting: transmitting radar waves between layers of air so that the waves travel farther than normal.

electromagnetic wave: a wave of energy produced by a combination of electricity and magnetism. Radio waves are one form of electromagnetic wave.

frequency: the number of energy waves that pass a point per second.

ionosphere: a layer of gas that surrounds the earth near the top of the atmosphere. The ionosphere reflects radio waves.

microbursts: small columns of air that travel toward the ground at speeds of up to 150 miles per hour.

modulator: the on-off switch in the radar set. The modulator tells the transmitter when to send radar waves to the antenna and when to shut off.

oscillator: the part of the radar set that generates electronic waves for the radar beam.

piloting: navigating a ship by following recognizable signs along the coast or in waters close to land.

radar: a device that locates distant objects by transmitting radio waves toward them and listening for reflected echoes.

radio waves: waves of energy produced by electricity and magnetism.

receiver: the part of the radar set that amplifies weak reflected radar signals picked up by the antenna. The receiver also filters out background noises that are picked up by the antenna.

signal processor: the part of the radar set that screens out echoes from large fixed objects, such as trees or mountains. It locates the desired reflected signals from among all the reflected radar waves it receives.

transmitter: the part of the radar set that amplifies energy waves from the oscillator and transforms them into high-powered electromagnetic waves. The transmitter generates short pulses of radio waves that are sent out by the antenna.

stealth aircraft: planes that are designed to be invisible to radar. This is accomplished both through the types of materials used to construct the plane and the shape of the plane.

trajectory: the curved path that a rocket or missile follows from launch until it reaches its target.

For Further Reading

■ ■

Above and Beyond, The Encyclopedia of Aviation and Space Science. Chicago: New Horizons Publishers, Inc., 1968.

Louis Battan, *Radar Observes the Weather.* Garden City, New York: Anchor Books, 1962.

Roy Braybrook, *The Aircraft Encyclopedia.* New York: Julian Messner, 1985.

Irwin Math, *Morse, Marconi and You.* New York: Charles Scribner's and Sons, 1979.

The New How It Works. Westport, Connecticut: H.S. Stuttman, Publishers, 1987.

Works Consulted

∎ ∎

James E. Brittain, "The Magnetron and the Beginnings of the Microwave Age," *Physics Today*, vol. 38, July 1985.

Michael Collins, *Liftoff, The Story of America's Adventure in Space*. New York: Grove Press, 1988.

Richard Collins, "Grand Canyon Collision: It Triggered Modern Air Traffic Control," *Flying*, vol. 114, September 1987.

Stewart Cowley, *Space Flight*. New York: Warwick Press, 1982.

Orrin Dunlap, Jr. *Radio's 100 Men of Science*. New York: Books for Libraries Press, 1970.

Edward Herron, *Miracle of the Air Waves*. New York: Julian Messner, 1969.

Brian Johnson, *The Secret War*. New York: Methuen, Inc., 1978.

William D. Marbach and Jeffrey Phillips, "New Tools for an Ancient Dig; a High-Tech Drill and Radar Unveil a Pharaoh's Boat," *Newsweek*, vol. 110, November 2, 1987.

Norman Polmar, *The Ships and Aircraft of the U.S. Fleet*. Annapolis: Naval Institute Press, 1987.

Stanley Wellborn, "Science Takes on Tornadoes," *U.S. News & World Report*, vol. 100, May 12, 1986.

Roberta Yared, "Jungle Radar," *Space World*, vol. W-12-276, December 1986.

Index

About the Author

▪ ▪

Deborah Hitzeroth currently lives in San Diego where she is working on her Master of English degree. Before moving to California, she lived in Missouri where she obtained a Bachelor of Journalism degree from the University of Missouri and an aversion to snow. After receiving her degree, she left the Ozark Mountains for sunny Texas where she spent three years working on a daily paper. Following a brief stint as a section editor on a weekly paper, she worked as a free-lance writer for a monthly medical magazine in New York. This is her first book-length manuscript.

Her husband, a naval officer and a certified space and science addict, helped spark her interest in radar.

Picture Credits

■ ■

Cover photo:
AP/ Wide World Photos, 12, 13, 19, 20, 24, 25, 26, 32, 33, 34, 47, 53, 56, 57, 58, 60, 61, 62, 64, 65, 71, 82
The Bettmann Archive, 15
Boeing, 83
Department of Defense, 81
Donna Eaton/Federal Aviation Administration, 48
Federal Aviation Administration, 50
Hughes Aircraft, 35
Library of Congress, 14, 18, 21, 22, 36, 37, 43, 44, 50
National Aeronautics and Space Administration, 59, 70, 74, 77, 78, 79, 80
National Center for Atmospheric Research/ National Science Foundation, 17, 25, 29, 31, 48, 52, 57
National Optical Astronomy Observatories, 71, 74, 76, 77
National Radio Astronomy Observatory, 27, 72, 73
The Port Authority of New York & New Jersey, 60
Raytheon Company, 38, 84
Lance Strozier/Federal Aviation Administration, 46
Uniphoto Picture Agency, 66, 68
U.S. Air Force, 38, 41
U.S. Army, 39
U.S. Navy, 42
Jim Wilson/National Center for Atmospheric Research, 51, 54, 56